BIRD DROPPINGS

Writings About Watching Birds and Bird Watchers

PETE DUNNE

featuring drawings by David Gothard
and a foreword by Ted Floyd

STACKPOLE BOOKS

Published by
STACKPOLE BOOKS
5067 Ritter Road
Mechanicsburg, PA 17055
www.stackpolebooks.com

Printed in the United States of America

10 9 8 7 6 5 4 3 2 1

First edition

The foreword in this book is excerpted from an essay that originally appeared in *Pete Dunne's New Jersey*, published by New Jersey Audubon in 2014.

Cover photo by Charles Brutlag/Shutterstock.com
Cover design by Tessa J. Sweigert

Cataloging-in-Publication Data is on file with the Library of Congress.

To Judy Toups, who inspired us all

Contents

Foreword: The Literary Style and Legacy of Pete Dunne

Pete Dunne is widely recognized as "The Bard of Birding." That epithet appears to have originated with a personality profile in the *Wall Street Journal*, and, inasmuch as it implies praise, the acclaim is well deserved. The epithet is also misleading, or at least insufficient as a descriptor of Dunne's literary style and legacy.

Pete Dunne's persona—at the podium as well as on the printed page—is charming and ingratiating. Folksy, even. It's also an act, a deliberate act, and I intend that as a compliment.

Like so many birders of my generation, I was introduced to Pete Dunne by way of *Tales of a Low-Rent Birder*. The storytelling was spellbinding. Thirty years later, it still is. My first real-life encounter with Dunne was in Arizona, in the early 1990s; I watched as he spun yarns for a bewitched gathering of teen birders at Camp Chiricahua. And nearly two decades after that, at a modest bird festival in a small town on the plains of southeastern Colorado, I marveled at Dunne's ability to captivate a crowd of barely-birders.

Charming. Ingratiating. Folksy.

It's a shtick. It's how Dunne gets through to audiences who might not otherwise apprehend the depth of feeling, the density of fact, and the "deep ecology" of his message. And it is, in some sense, a pity, for it has obscured, I believe, the literary merit of Pete Dunne's oeuvre.

At the New Jersey Audubon Society's "Celebration of Pete Dunne," held in June 2014, numerous birding luminaries streamed to the dais for the honor of reminiscing about the life and legacy of Pete Dunne. The tributes were largely personal—with one notable exception. Dunne's longtime friend, the legendary and iconoclastic

Kenn Kaufman, bucked the trend. Kaufman put the audience on alert that he would be speaking of the ways in which Dunne has influenced and inspired all the people who *don't* know him.

Like me, for so many years.

True, Pete Dunne has become a personal friend. But he was an influence and an inspiration long before I ever met him. How did that happen?

The initial hook, no doubt, was the charm; it flies off the pages of Dunne's books and essays. The enduring appeal, I flatter myself to imagine, has been the depth and breadth of Dunne's worldview. But there's something else, and Kaufman came right out with it: Pete Dunne is a great writer. According to Kaufman: Dunne's sentences are short, his words familiar; his prose is direct and declarative; he gets to the point.

Kaufman put an idea in my head, and I sent an email to Dunne. I got this response:

> I think Hemingway is the greatest writer of the twentieth century, maybe of all time. I honestly don't know how much influence he has had upon my writing. I admire his writing but have never tried to emulate it. Every writer needs to find their own voice.

I confess, I was a late bloomer when it comes to Ernest Hemingway. I blame that in part on the educational system. I was *made* to read Hemingway, and what self-respecting, know-it-all teenager would ever assent to approve of any assigned reading? Hemingway's no-nonsense writing—call it "direct and declarative," and characterized by short sentences—was simple and artless, it seemed to me.

I've come to my senses about Hemingway. And I've come to appreciate that Pete Dunne is the Ernest Hemingway of American nature writing—even though I would have never, ever, imagined such a thing at the time of my first contact with Dunne's writings. Why, Pete Dunne is a nature writer, I would have insisted.

Contemporary American nature writing is many things: sensitive and sprawling, "Eastern" and perhaps feminine, holistic and allusive, lyrical, and more. Hemingway's muscular and masculine writing tends not to be those things.

Every writer needs to find their own voice.

Any contemporary nature writer who recalls Hemingway has found his own voice. Mind you, I'm not talking about Hemingway's hallmark themes: war, romance, maleness, etc. Rather, I refer to Hemingway's direct and economical writing.

Case in point: "Embers of Spring," which I first encountered years ago in Dunne's *New York Times* column, "In the Natural State." The essay is all of 421 words, yet "Embers of Spring" is landscape art. "Embers of Spring" also suggests a haiku, the way it evokes multiple natural images at the same time: logs slowly burning on a fireplace, a Red-winged Blackbird singing for an instant, and the grand pageantry of springtime.

"Embers of Spring" is dense, in a good way, like Annie Dillard's re-creation of a snowstorm; it is purely descriptive, in a wondrous way, like Henry Beston's nor'easter; it is indelible, like the haiku of Allan Burns. Yet one does not ordinarily find "Embers of Spring"— nor any of Dunne's writings, so far as I know—on the "literature" shelves in bookstores and libraries.

Not yet.

Deep ecology, as I have come to understand the term, is a realist philosophy that seeks to unencumber environmental awareness from a human agenda. And deep ecology pervades many of Dunne's greatest essays: "First-Year Bird," a nihilistic entry in *Tales of a Low-Rent Birder;* "Golden Plover at Ebb Tide," the elegiac yet dispassionate opening chapter of *More Tales of a Low-Rent Birder;* and practically everything in *Before the Echo.* Even in Dunne's unabashedly utilitarian books—*Hawks in Flight, Pete Dunne on Bird Watching,* and *Pete Dunne's Essential Field Guide Companion* —there is, I sense, an undercurrent of deep ecology.

Does that not seem odd? For one thing, Pete Dunne is, somewhat controversially, a hunter. He's also one of birding's greatest ambassadors. He's "The Bard of Birding." Why, in my remarks alongside Kenn Kaufman's in New Jersey, I acknowledged Dunne for his contributions to "birding, conservation, [and] humanity."

Especially his humanity. I noted at the outset that Dunne is charming and ingratiating; he's folksy. He's a stare-you-in-the-eye, shake-your-hand-hard, all-around good guy. Can an authentic "deep ecologist"—can any self-proclaimed environmentalist—be possessed of so sanguine an outlook on life?

My answer to that question, until recently, would have been in the negative. I think of the anger in Terry Tempest Williams' *Refuge*, that book which provides the closest "comp" to the kaleidoscopic rhetoric of Dunne's astonishing "Small-Headed Flycatcher. Seen Yesterday. He Didn't Leave His Name." I think of my friend Allan Burns; even though his haiku are pure art, Burns is, in real life, a sharp and principled critic of much about the American way of life. Most of all, I think of the radical, Aldo Leopold, whose literary land-scape art—for example, "Guacamaja," "Too Early," and "On a Mon-ument to the Pigeon"—so clearly anticipates Dunne's themes and images. Many have tried to soften Leopold's image, but the effort is in vain. *A Sand County Almanac* opens with a broadside against "our Abrahamic concept of land" and ends (in an expanded compi-lation) on a note of despair about "the still unlovely human mind."

I've recently had to re-examine my assumptions.

Jeffrey A. Gordon, president of the American Birding Associa-tion, and I have disagreed over the years on various matters, including the "proper" attitude of the environmentalist. I've long maintained, I've always assumed, that the eco-greats—from Henry David Thoreau to Paul Ehrlich, from Aldo Leopold to Terry Tempest Williams—are perforce resentful of, and at times openly hostile to, prevailing norms in contemporary culture and society. You actually *cannot*, I would have told you, be an environmentalist without having a dark view of humanity. Not of individual humans, mind you, but rather of humanity, of our collective imprint on this Earth.

I have to say, Jeff Gordon has nudged me toward a revised conception of the environmentalist spirit.

Thoreau and his literary and intellectual heirs left their mark on the environmental movements of the nineteenth and twentieth centuries. I'll never back down from that viewpoint. But I think Jeff's onto something about the "green" worldview and way of life in the early twenty-first century. And I think Pete Dunne's literary style and legacy have particular relevance to the contemporary environmental movement.

Pete Dunne's writing, at its very best, has untied the Gordian knot of environmentalism—in its normative guise, an odd admixture of progressivism and despair. Dunne's influence, whether it is direct or indirect, is undeniably present in the output of today's most prolific young birders: Noah Strycker, Tom Johnson, Jen Brumfield, and others. In their writings and other communications, they have embarked on a sort of secular *via positiva*, the same spirit that fills the pages of Dunne's writings.

There will never be another *Tales of a Low-Rent Birder*. Pete Dunne's agent has told me so! Yet it has paved the way, I believe, to idle reimagining of the modern environmentalist movement: more engaged and affirming, inclusive and holistic, absurd at times and able to laugh at itself, yet uncompromising in its deep ecology.

There's one more thing.

I hope Pete Dunne's literary legacy translates into successes for the environmental movement in the early twenty-first century and beyond. I do hope so. But I'm not sure. Of one thing, though, I am nearly certain.

The time will surely come when Pete Dunne's greatest "essays"— except that's not quite the right word, for I still don't know their genre—will go on the same shelf as Terry Tempest Williams and Allan Burns, even Henry David Thoreau and Aldo Leopold.

And Ernest Hemingway.

—Ted Floyd
editor of *Birding*

Introduction

Half a lifetime ago I was brought to ask the prolific New Jersey writer/historian John T. Cunningham, "How many articles have you written?" His response has long stayed with me.

The historian laughed. Long and hard. Then, in response to the chagrin and puzzlement that must have mapped the face of the aspiring young writer before him, he said, "I truly have no idea."

At the time, I found it difficult to conceive that any writer, no matter how prolific, could lose track of their work and be so cavalier about it. Now, many thousands of articles and essays into a writing career that spans five decades, I appreciate John's mirth and perspective.

I too can no longer quantify my writings. I doubt I can even recall every column I've penned under my name (or another).

This collection of mostly short essays, or "droppings," many originally written for my column "The Catbird Seat," featured in the Cornell Lab of Ornithology's *Living Bird*, are personal favorites—the sort that linger in memory. Others that have earned a place in this collection were drawn from another favorite repository: a column entitled "Birder at Large," first written for the old *Audubon's American Birds* and later recast in *Birder's World* (now *BirdWatching*). Also included here you will find a few essays drawn from "Beak to Tale," my once quarterly offering to readers of the newsletter of Wild Bird Centers of America, and from "Bird Droppings," a column I pen for a freewheeling Cape May tabloid called *Exit Zero*.

Why is it that some essays write better than others? Often it is because the editors—including, and most particularly, Tim Gallagher of *Living Bird*, Chuck Hagner of *BirdWatching*, and Jack Wright of *Exit Zero*—give their writers lots of editorial license. Writers write best when they are having fun, and I don't recall a time when penning pieces for these publications was anything less.

Among these essays you will find many that poke a little good-natured fun at our avocation and its practitioners. Birding, after all, is an activity that is supposed to be fun.

This is not to say that this collection does not include articles tailored to be informative, helpful, even provocative. For instance, have you ever wondered what to do with that drawer filled with retired optics—the ones you purchased and discarded before finally acquiring the quality instrument you wanted all along? In these pages, I'll tell you.

Have you ever mistaken a lawn flamingo for a real flamingo? Probably not. But the species account treating the identification of Lawn Flamingo may help you avoid making this faux pas.

Hopefully, it will also make you laugh.

This is a great time to be a person who enjoys watching birds. As retired director of the Cape May Bird Observatory, I have had, for more than forty years, the privilege of watching our avocation grow and mature. And while these essays do not claim to have changed the face of birding in North America, I believe they do capture something of the essence of it, and point out some of our foibles, too. I hope you enjoy them.

Writing, as I love to point out, is fifty percent reader. We're a team, you and me. What you bring to the reading experience will make each of the essays in this collection unique. For this, you have my respect and gratitude.

Pete Dunne
Cape May, New Jersey
Autumn 2014

Why There Are Birds

"Uncle Peter, why are there birds?"

"Excuse me?" I said, turning, looking down into a face so steeped in innocence that Shirley Temple would have traded up.

"I said," intoned the little cherub (her face now contorted by a beguiling pout), "why are there birds?"

To which I replied . . .

"Duhhhhh . . ."

I'm not often at a loss for words. But the kid stopped me. Been studying birds for fifty-seven years, never occurred to me to examine the reason for their existence. Yet there seemed no way to avoid a question so honestly asked.

"Well," (I considered saying) "birds are here to keep the farmers' crops from being consumed by harmful insect pests." It was a half-truth I'd exhumed from some grammar school science text. This widely accepted assertion, first spun by agricultural agents a century ago and widely embraced by the conservation community, was usually supported by speculative estimates of grasshopper reproductive capacity, which, given one hundred percent survival and three or four generations, would result in the earth's land mass being covered with a six-foot-thick sheet of crop-munching insects.

But birds were here long before our species tried its hand at farming. It hardly seems plausible that birds were put here in anticipation of the advent of agriculture. So I considered other possibilities.

"Well," (I started to say) "birds are here to fill an important ecological niche—in fact, niches." As both predators and prey, birds are important components of the food chain and are therefore key players in the balance of life on earth.

But one look at those innocent eyes disclosed the folly of this discourse. Who wants to be the first to tell a six-year-old that baby robin is on the menu?

"Birds are here," (I almost said) "because without birds there would be no nonstop flights to Orlando." It's unlikely our earthbound species would have grasped the potential of an airfoil without seeing birds in flight. And could civilization as we know it exist without an air link to the Magic Kingdom?

"OK," (I semi-postulated) "birds are here because without them a lot of professional football franchises wouldn't have names." Imagine what would happen to season ticket sales if fans were entreated to root for the Philadelphia No Names or the Baltimore No Such Animal or the Seattle Fill in the Blank.

And literature! Consider how the absence of birds would change the face of literature. Do you think a book entitled "To Kill a Marsupial" could have made the *New York Times* best-seller list? Would Edgar Allan Poe be remembered if the line read: "Quoth the Thirteen-lined Ground Squirrel, 'Nevermore'"?

I was beginning to get desperate. The attention span of the average six-year-old lasts about as long as innocence. I was frantic to come up with some answer this spawn of suburbia could accept.

"Lawn care products," (I almost shouted). "No advertising agency worth its guile would consider airing an ad promising a richer, greener lawn without a Mourning Warbler or Savannah Sparrow singing in the background.

"I'll bet you like helping your daddy cut the grass?" I encouraged.

"Uh uh!" she said, screwing up her face the way six-year-olds do when grown-ups say something really stupid.

I knew I was facing failure. The Uncle Who Knew Everything about Everything was stumped. My mind raced. I grasped at straws.

There are birds because . . . because . . .

Without birds, cats would have no reason to sit on windowsills . . .

Car washes would go broke . . .

Bad little boys would stop dreaming of getting BB guns for Christmas . . .

Such a simple question. So elusive an answer. So . . .

I just came clean. Told my niece the truth as I know it. When all else fails, try honesty.

"There are birds, sweetie, because without birds your Uncle Peter wouldn't have a job."

"Oh," she said. Studying the ground. Pursing her lips. Nodding sagely.

"Uncle Peter, why are there wars?"

I don't know this for certain, but I do believe that if world leaders were to spend just one hour a day watching birds, there would be fewer wars to trouble the minds of six-year-olds.

Gift of Chickadee

Somewhere between the tumble out of bed and the morning commute, a ritual conducted with the taste of toothpaste still in my mouth and the anticipation of brewed coffee ahead, comes the greatest moment of the day: the moment I fill our bird feeders (my daily offering to the universe) and receive the gift of chickadee (the universe's return gift to me).

It is a thing I have done since childhood, a ritual as focused as Zen and as expansive as prayer. Each morning, I step into an ordinary suburban yard. I ladle humble offerings into a vessel studded with perches. Then, grasping the plastic urn in both hands, I raise it to some lofty hook, step back, and . . .

Am rewarded by the gift of chickadees (or titmice . . . or nuthatches . . . or whatever the universe is serving up this morning). Usually it is chickadees. What species? It hardly matters, but for the record, for this period in my life's journey, it's Carolina Chickadee. In my youth, in North Jersey, the envoys were all Black-cappeds.

What is so special about this ritual? Everything. All the small things. All the interlocking elements that are gratifying in their sameness and exciting in the possibilities they unlock. I like:

The cold that wraps itself around me like a penitent's smock, giving me kinship and standing with birds who have faced down the trials of the night.

I like the wholesome smell of seed that makes me bring my nose close, inhaling deeply.

I like the *s t r e t c h*. When I battle the forces of gravity (and aging joints, and clothes that constrict more than they did before the holidays) and feel the gratifying tug of the feeder's handle finding its hook, I'm assured that I have met and mastered my first challenge of the day, giving me courage to confront the rest: internet service that won't connect, coffeemakers that throw a tantrum all over the counter, printers that run out of toner.

And of course I love the moment I step back, arms to my sides, anticipation hardly in check, and see the tiny forms materialize out of the half-light, taking their perches, perhaps honoring me with a glance or, turning their backs, showing me that in this world there is still grounds for trust between living things.

OK, I'll grant that filling bird feeders is not exactly a pivotal point in the sweeping drama that is the human story. It is not as significant as, say, the invention of the wheel, the signing of the Magna Carta, the splitting of the atom, or even a sale on ground chuck at the supermarket.

It doesn't matter that wild birds, being wild birds, can get along without my offering, or that what I presume to call "gratitude" might more accurately be called "tolerance," even "indifference."

I'm not looking for gratitude. Like the biblical brothers Cain and Abel, all I'm looking for is confirmation. The hungry chickadees, coming in from a smoke-colored dawn, give me this, wrapped in the blessing of their trust.

It is a complex world we live in, and far from an ideal one. There is good; there is evil. There is too much need and too little peace and too many hurtful things I would like to set right, if only I had the wisdom, and if only I were given the chance.

But I lack the former and it seems likely I'll not be granted the latter. One thing I know. Every day I rise, there is a feeder that I can fill and there are chickadees that anticipate my offering, and this small interaction between living things is good—as good as good gives and gets.

The Order of Birds

"This is stupid," a member of my beginning birding class asserted.

"This" covers a wide range of subjects. Insofar as we were sitting in a classroom, doing the ritual checklist for the birds we'd encountered on our morning outing, several interpretations came to mind.

Was he perhaps reflecting upon the action? The act of going through a list of birds and checking off the ones seen? Silly it might be. But stupid?

Was he passing judgment on the quality of the list (which was, I admit, printed on some pretty cheap paper)?

Was he upset because there wasn't enough light or he'd forgotten his reading glasses or . . . ?

"Could you be a bit more vague?" I invited.

"This," he said, "holding up the list. These names. The order. Why are they in such a stupid order? You can't find anything."

"Ahhh," I said, finally getting the point. This was the gentleman's first encounter with the AOU Checklist, which arranges birds not alphabetically but in their taxonomic order—from most evolutionarily primitive to most advanced.

"But you *can* find things," I insisted. "All you need to do is learn the order."

"What's the matter with the alphabet?" he wanted to know.

"Because then birds wouldn't be ordered and grouped according to their similarities. It would be like going into the grocery store and finding anchovies next to avocados, bananas flanked by bacon and biscuits."

"The dictionary is in alphabetical order," he said (not giving an inch). "Does anybody get upset because two words that have nothing in common follow each other in a dictionary?"

I had to admit he had a point. Told him so. And while I hoped this would mollify him (allowing us to get on with the checklist), I was wrong.

"Just name me one other ordering of things where the letters don't follow each other in alphabetical order," he demanded (the "and I'll concede the point" was implied, or maybe just presumed).

And I smiled. Smiled as Oedipus might have smiled after hearing out the Sphinx. But I figured I'd save and savor my verbal coup de grâce for a moment. I wanted to contemplate and admire the utilitarian order of the checklist first.

The AOU Checklist—the checklist of North American birds as formulated by the members of the Committee on Classification and Nomenclature of the American Ornithologists' Union. It places gulls in front of terns. It places shorebirds ahead of gulls and plovers ahead of sandpipers.

In the studied estimate of the American Ornithologists' Union, loons are more primitive than Wild Turkeys (even though loons can swim and dive and turkeys cannot). And the most advanced species of them all is, believe it or not, the common House Sparrow.

No, I don't know why. I don't sit on the committee.

But my point is that the AOU Checklist is not evil. It doesn't challenge or undermine the organizational stranglehold enjoyed these many years by the Roman alphabet. Its organization makes perfect sense when the purpose is to arrange a biological train on a continental dance floor with birds that have little in common spaced well apart in the ranking, and those that are similar proximally linked.

This is not to say that the order is immutable or not subject to change. Take loons. For years the AOU Checklist began with the Red-throated Loon. You don't see them in the front of the checklist anymore. They've been reassigned to a slot following waterfowl. A change motivated by the latest scientific rethink. For the foreseeable future, the checklist will begin not with the Red-throated Loon but with the Black-bellied Whistling-Duck.

Get used to it. You got used to gas at four dollars a gallon (and you will again).

But back to the gentleman's challenge.

"You name me one ordering of things where letters don't follow each other in alphabetical order."

"The periodic table," I suggested.

"OK, name me one other," he invited.

"The days of the week?" I offered.

Some Girl Named Cindy

We passed on the elevated loop above the swamp, mouthing greetings with averted eyes (as men and women who are strangers and unexpectedly find themselves alone and in each other's company do these days). The second time we passed, we exchanged nods, but by the third round she honored me with her voice.

"I think that this is pretty special," she said.

"This place?" I asked.

"The sound. The male alligators roaring. I don't think it happens every morning."

"I wouldn't know," I replied, "not being a frequent visitor to the Everglades."

"I'm not either," she confided. "This is only my second time. The first was on a college trip."

About twenty years ago, I guessed, looking at her face, her eyes, the set of her mouth. It was a pretty face.

"So what good fortune got you to sunny here and away from a cold where?"

"New Hampshire," she said. "Got laid off," she added with a shrug. "Had some frequent flier miles that I had to use or lose. I thought, why not the Everglades?"

"Good for you. I aspire to unemployment myself," I replied, trying to put an upbeat spin on her disclosure.

"It stops being fun after a year and the bills start piling up," she said, locking eyes. And I guessed, plumbing the depths of those eyes, that there were things not said—a recent death, perhaps, or a marriage gone sour. It's been my experience that people whose lives are touched by loss or sadness gravitate to the ends of peninsulas to take stock and maybe find new beginnings. When people reach these literal ends of the earth, their footsteps seem inexorably guided to visit natural areas.

"Did you see the young Anhingas?" I invited, to change the subject, to give a troubled soul the only thing I had to give, which was my knowledge and my insight.

"No," she admitted. "Where?"

"There," I said. Setting up my spotting scope. Training it on the slender powder puffs in the nest. Inviting her gaze.

She looked through the scope and her mouth widened into a grin her college self would have been challenged to match.

"Oh, they're beautiful!" she exclaimed. "How did you ever find them?"

"I've been watching birds a long time," I said. "By the way, my name is Pete."

"Cindy," she said. "Are there other birds around?"

We navigated the boardwalk for half an hour. She, seeing birds such as Least Bittern and Boat-tailed Grackle for the first time. Me, through her eyes, seeing them again for the first time. When we parted—me to find wife Linda, Cindy to continue her search—we parted with smiles. I didn't expect to see her again. I was wrong.

On my way to our car, I was hailed by a voice from the Everglades Park gift-shop/bookstore.

"Pete," she said. "Just one more question. I want to know which field guide to buy. There are so many."

"Peterson's *Eastern*," I shouted automatically.

"Who was that?" Linda wanted to know when we were seated in the car.

"Some girl named Cindy. From New Hampshire. New birder. That's all I know."

The rest is conjecture.

What's Your Favorite?

"**A**ny more questions?" I invited.

"No," the most recent in a long string of reporters replied. "That about wraps it up," he said, snapping his notepad closed for effect. "Oh," he said, reopening it. "One last thing."

Here it comes, I thought, and as he enunciated the question, I formed it in my mind.

"What," he said (and I thought), "is your favorite bird?"

He hadn't made it the first question in the interview. That's how I knew it would be the last.

"Favorite bird?" I repeated, confronting once again my least-favorite question. "Favorite bird?" I said again, dredging up all the usual replies, considering the merits of each one.

Sometimes (if I'm bored with the interview) I say "Northern Harrier," which is close enough to the truth. Fact is, I think harriers are extraordinary birds. Beautiful, nimble, and unlike many raptors that sit and wait for fortune to favor them, it's a species that really hunts. I've been fascinated by harriers for thirty years, seen them do amazing things (including steal prey from a Peregrine Falcon), but . . .

Honesty compels me to admit that Swallow-tailed Kite, not Northern Harrier, is my poetic favorite. There isn't a birder alive who could argue against the choice of Swallow-tailed Kite for title of Favorite Bird, but . . .

To choose Swallow-tailed Kite would mean precluding any jays (including Blue Jay—a favorite bird of my youth). At times bold and brash, yet shy and sly, too, Blue Jay are the Tom Sawyers of the bird world. Screaming blue murder one minute, murmuring apologies the next. The sight of a Blue Jay on a winter day makes me want to cheer, to pin the label "My Favorite Bird" on its tail, except . . .

Black-capped Chickadee would then be runner-up. And nobody who loves chickadees as much as I do could abide seeing these friendly acrobats diminished by a second-class standing. Favorite bird? Favorite bird. Favorite . . . vocalist? That would be Winter Wren. Sentimental favorite? That would be Wood Thrush. Favorite one-time encounter with an individual bird? That would be American Robin. Name was Chipper. (You'll read about that encounter later in this collection.)

"You seem to be having trouble with the question," the reporter said.

"With nearly ten thousand possibilities, it's not an easy question to answer."

"That's OK," he said. "I've got plenty of material for the story."

He got up to leave. He started to leave. I saw motion in the bushes just over his shoulder—a splash of light, then color and a telltale vibration in the leaves.

"Wait," I said, pointing. "*That's* my favorite bird."

"Where?" he said.

"There," I noted.

"What is it?" he asked.

"I don't know," I replied to uncomprehending eyes. "But it's the bird in front of me right now. That makes it my favorite."

"Oh," he said, not bothering to open his notebook, not inclined to consider my observation copy worthy.

"Can I call you if I have any follow-up questions?"

"Sure," I said. "Good luck with the story."

"Thanks," he said, moving away, not lingering to see my favorite bird in the whole world.

If he wasn't really interested, I wonder why he asked?

Letter to Students Returning to Fall Classes, with Condolences from a Recovering Adult

You're right. It stinks. Life was made for summer. So why is it that summer is only three months long and the school year lasts an eternity? It is precisely to prepare you for the long, dull, pedantic road ahead called adulthood.

In a just universe, one governed by an intelligence that truly cared, all the delicious fun and discovery and ice cream and freedom and going to bed late that is summer would go on forever. There really is no defensible reason why all life's good stuff got crammed into a corner of the calendar year, or why it has to end.

Heck, fun doesn't take up any room. It's earth friendly. You can share it. And about the only place it doesn't thrive is school. So why this monstrous injustice?

In a word: *society.* If you and your fellow students ever suspected that you could simply have fun all the time, there would be no leveraging incentive for you to become the mindless, regimented, productive members of society that education strives to forge. You don't really believe that we taxpayers are picking up the tab for your education just to see you achieve your innate potential, do you? Only first-year teachers and guidance counselors believe that.

Me? I loathed school. Started getting the end-of-summer blues on July 4th. But by the time I was ten, I'd figured out how to get off the treadmill and make summer last through the school year, and how to neutralize all the onerous Right Think that was being

packaged and served up to me in subject-sized portions (yes, I'm talking about algebra, social studies, English comp, chemistry). Here's an example: Avogadro's number—6.02 x 10 to the 23rd moles per liter. I have no idea what this means. Until this essay, I've never even called this bit of useless information up in my mind, yet for one whole year in high school, every chemistry quiz began with the question "What is Avogadro's number?"

I always got it right, but I never did understand what it was or what significance it holds. I'm prepared to say at this stage of my life: none.

What, you ask, was the source of my epiphany? My realization that summers can be eternal, and that school is a mind-numbing trap?

I discovered birds! Fun to find. Fun to watch. Fun to share. More animate, real, and engaging than mathematical theorems, diagramming sentences, or, going back even further, reading about the values-instilling escapades of Dick, Jane, and Sally.

Who were Dick, Jane, and Sally? Kool-Aid drinkers, fictionalized role models we were supposed to bond with and relate to. They went on to invent the internet, engineer the stock market crash, perpetuate the burning of fossil fuels, and squander their lives in the futile pursuit of happiness.

One thing they didn't do was become bird watchers. I know this because if they had, they would have realized that happiness isn't to be pursued, it is to be enjoyed. Like life.

Listen kid. Summer isn't over, nor is fun, it is simply being hijacked by institutionalized stupidity. My advice to you, my erudite young pawn, is to start watching birds. They're fun. They're free. They make great role models. They put you on the road toward a life of fun that knows no seasonal or geographic or institutional boundaries. Not only that, you don't even need to pursue them. Happily, over most of the planet, birds come to us.

Look. Out the window. There's one now. On the grass, just this side of the teacher's parking lot.

Feel better already, don't you?

Are We Weird (Or What)?

Don't begin to tell me you haven't asked yourself this question —wondered whether the stigma that goes with being one of those kinds of people doesn't, maybe, have some foundation. What kind of person? You know. Go ahead and say it. Say it out loud.

"Bird watcher."

Kind of makes you wince, doesn't it? It's why we changed the name to "birder" a while back. "Birder" sounds more turbocharged. More in semantic accord with labels like golfer, hunter, shopper, stripper, mugger (and other activities society takes in comfortable stride).

You've seen the reactions of colleagues when they ask you what you did over the weekend and you say you chased a Bar-tailed Godwit. "That's a shorebird with a really long bill," you explain— and they change the subject to their golf handicaps or the war on crabgrass.

You've noticed how your kids hustle their friends out of the room when you blunder in wearing binoculars and a baggy smock with an uncountable number of pockets.

Yeah, you know.

So let's be honest. Let's say that there *is* something to other people's deprecating regard. Let's admit that being a bird watcher has a quirky side. Admit . . .

That going to the lunch counter in a midsize western town and asking whether anyone can give you directions to the sewage treatment plant is not exactly mainstream behavior.

Or that you would be hard-pressed to explain why you were eager (not just willing) to spend thousands of dollars to fly to an island that is so far west they had to move the International Date Line to keep it from being east, all because some bird that doesn't belong there might appear and then you get to count it on a list because according to birding's rules and regs, now it's in bounds.

I'm not saying that birding is unique in its quirkiness. After all, society's rank and file have their share of quirky affectations, too.

Take lawns. I know people whose lives are indentured to the task of maintaining a vegetative monoculture whose sole purpose is to lie flat and look green. While this might sound silly, it is also normal.

Take football. Don't you think it's curious that America's Sunday afternoon passion is watching two herds of mesomorphs physically maneuver a ball up and down a flat grid all because the sphere is so misshapen it can't roll the length of the field by itself?

Curious this fascination might be. But nobody thinks it isn't normal.

Consider slot machines. Now here's this machine whose sole purpose is to relieve people of their money, but it's designed to do it slowly so the hemorrhaging can go on for hours. I've never met a person who admitted they like to lose money. But there are plenty of people who like to gamble.

Back to birding. I was a bird watcher even before there were birders, and over time I've grown inured to both the subtle and not-so-subtle slights that come with the avocational turf. But that doesn't mean I haven't been wounded, and it doesn't mean that part of me hasn't at times wondered whether birding (or bird watching) might actually *not* fall within the bounds of normal behavior and that I therefore am not normal. I just never had anyone to ask (since all my friends are birders).

Then, just this year, I got my chance. I was staying in a hotel where a bird-watching convention was being held. There, behind the front desk, was a normal human being. You could tell this because she wasn't wearing binoculars, she seemed not to know where the sewage treatment ponds were, and all her nails were painted like multicolored cupcakes. So I asked her.

"So . . . what do you think of birders? A little weird, maybe?"

Her response was not immediate. I'm inclined to think she gave her answer some thought.

"Well," she said, "they're mostly nice. And some of them seem pretty normal."

"Thanks," I said. "Nice nails. Must take a lot of time."

"Hours," she admitted.

WARNING: You Are Entering Birder Country

WARNING: You are entering an area known to be infested with birders. For your safety and theirs, please read and heed the following instructions. It may save you considerable distress.

Despite their well-earned reputation for odd behavior, birders tend to be fairly docile and are commonly not dangerous unless provoked. Knowing this, and learning how to behave when encountering a birder, will make you more secure and greatly increase your chances of avoiding an embarrassing confrontation.

First, if you meet a birder in the field, don't appear startled. If a birder senses that you find their interest the least bit odd, they will almost certainly respond with focused indifference.

Second, look friendly, even interested. Birders respond to honest, intelligent questions the way tent-revival ministers embrace sinners. The proper way to express your interest is to ask "Seen anything good?" or "What have you got out there?"

Never try to be funny or clever. Do not, for example, walk up to a birder scanning distant sandbars and ask "See any Ivory-billed Woodpeckers out there?"

You, of course, are simply trying to express your interest and awareness of current events. But a birder is more likely to respond to your question by thinking, "What kind of a mental midget thinks that a forest woodpecker would be dog-paddling around in an ocean?"

They might not say this. But they will certainly think it. And having now distinguished yourself as an idiot, your chances of surviving the encounter in an amicable fashion are diminished.

Be careful. Even things said in honest innocence might trigger a dismissive response from a birder. Things like "Are you looking at ships?" or "How far can you see with those things?" might seem unprovocative, even germane, to you. But to a birder feverishly scanning for a reported Black-tailed Gull, such oblique inquiries are likely to pose an unwanted distraction.

Watch for these danger signs: Eyes rolled skyward. Eyes squeezed shut and head moved slowly side to side. Head turned slowly your way and face frozen in an expression that suggests your nose was put on upside down.

This is a very dangerous situation. Often a birder will at this point hand you their binoculars and invite you to "see for yourself."

Their objective is to help you dig your own grave deeper and faster by demonstrating your ineptitude with binoculars. Now you have only one chance for redemption.

Appear bold and confident. Grasp the binoculars with both hands (to guard against dropping $2,500 worth of glass in the sand).

Do *not* put the strap around your neck. This denotes possession, and even a Gandhi among birders will fight to the death to retrieve a $2,500 pair of binoculars.

Raise the binoculars to your eyes and move the focus wheel (on a $2,500 instrument, it should be precisely where your index finger falls) until you achieve a clear image.

Pan the binocular across the horizon, stopping at one, maybe two, places en route (as though you were studying something).

Draw the binoculars away from your eyes. Give them an appraising look. Hand them back to the birder and say, "Nice."

Birders are intensely vain about their binoculars. Your recognition of the instrument's superior quality will not only address the birder's vanity but also assuage their repressed anxiety about buying such an expensive toy in the first place (since there are $300 binoculars out there that will do pretty much the same thing that the $2,500 models do).

Their vanity stroked, their anxiety diminished, you will find that birders become docile, even cordial. They may start making esoteric references to things like a such-and-such-and-such "back cross" or an "arrested second-cycle" something or other "gull." They may invite you to look at digiscope images of the bird they were "lucky enough to find yesterday."

Don't be fooled. This is still a wild birder, deserving of your utmost respect. One incautious action on your part can turn this now-friendly individual into a force of nature. If (for example) you fumble the handoff and $2,500 worth of binoculars falls into the sand, if your sudden movement causes the gulls loafing on the breakwater to flush and head for the horizon, if another birder calls out "I've got the bird over here" and you happen to be standing in the way—your chances of becoming collateral damage are high.

Don't let yourself become a statistic. Learn to respect birders and keep a safe distance. Remember, you have chosen to place yourself in a habitat where birders are common. For your safety, and the safety of other nonbirders, show them the respect they are due.

Of Birds, Van Gogh, and Cheap Beer

I am sometimes asked (most commonly by Martians and NASCAR fans) why anyone would want to waste their #$@%!& time doing something as stupid as watching birds. Not wishing to add to the burden of their lives, I quickly agree that bird watching certainly is frivolous and did they know that quart bottles of Schlitz were half price at the local package goods store?

I'll bet you didn't know Martians drink Schlitz.

They do when they can't get Old Milwaukee. Mars is a dry planet.

But sometimes inquiries relating to the appeal of bird watching are of a truly inquiring nature (you can usually tell this because the question is posed without expletives).

So why do millions of otherwise well-adjusted adults arrange the moments of their day just to engage birds?

"Well," I sometimes explain, "do you like going to art galleries? Birds are akin to fine works of art, and birds add an element to the viewing experience that a painting cannot. Birds have free will. They fly. So getting a satisfactory look at the object of your desire is more challenging and, in equal measure, more gratifying when you succeed.

"And," I point out, "you go to a gallery today, you go to the same gallery tomorrow, you see all the same paintings in all the same

23

places. Birds are different, always shifting—here today, gone tomorrow. But there is a good chance that a departed bird will be replaced by another species (maybe something unexpected—the equivalent of a rare stamp among birds or an undiscovered Picasso)."

Imagine going to a gallery and finding a Van Gogh, turning to study a Renoir, then returning to the Van Gogh only to discover that it has been replaced by a landscape by one of the Dutch masters. Which Dutch master? That's for you to determine.

Part of bird watching's appeal is the challenge of sleuthing. You have to *study* the bird. Note distinguishing characteristics. Match them to the illustration and text in your field guide (the bird-watching equivalent of a gallery catalog). Pin the name to the bird and you cast a spell of binding upon it. You make it yours, another feather in your cap.

Which leads to yet another appealing aspect of birding. You get to collect all the birds you find on a list. Once you've found and named all the birds close to home, you travel and find birds that are farther afield—in places like Florida and Texas, Alaska and California, Peru, Kenya, Australia, Antarctica. Ten thousand possibilities in all.

Hey, boys and girls.

Collect them all. Win a lifetime of travel.

And like so many things that motivate our species, birding involves the element of win and lose, gain and loss. When you see a bird you really aspire to see, you score! If you strive and fail, you lose.

And if you see a *really* good bird (an especially rare or beautiful bird), you get to share it with birding friends (invoking the social side of birding). Conversely, if you see a really special bird and your birding friends don't, you get to crow.

They *eat* crow.

Nah, nah, na, nah, nah.

Finding a rare bird is like getting a new car and parking it in the drive (except that seeing most of the planet's birds usually costs a good deal less than buying a luxury sedan).

Which brings up one of the best reasons to take up birding. It's cheep! Binoculars and a field guide are all you really need to get on the road to a lifetime of discovery and adventure.

With the money you'll save not going to the racetrack, you'll be able to upgrade your beer.

Field-Trip Guidelines

One of birding's most cherished traditions is the field trip. The ingredients include one leader, a variable number of leadees, a date and destination, lots of birds.

In a previous essay I offered a how-to blueprint for incipient field-trip leaders—a list of dos and don'ts that, if adhered to, would confer upon even a Walter Mitty among birders the semblance (if not the talent) of a Jon Dunn, Michael O'Brien, or David Wolf. It proved so successful that it caused an imbalance in the leader/leadee ratio to the point that ten newly minted (and self-appointed) field-trip leaders were turning up for every individual participant.

To right this imbalance, I would like to offer this complementary set of guidelines for field-trip participants. Follow them and not only will your standing as the alpha participant be assured, you will almost certainly get to see more birds than your fellow trip-goers.

1. Dress the part. An alpha participant always looks like a birder even when off duty (i.e., asleep). Check thrift-shop bins for used Patagonia and Mountain Hardwear gear. If the items are still too expensive, neatly excise the labels with a sharp knife and affix them to knockoff brands. Hats should be floppy, tasteless, and bedecked with no fewer than thirty bird pins. Jackets should be at least fifty percent covered by bird club and destination patches. Remember to leave room for the brand label.

2. Arrive early. Through conversation learn the leader's favorite birding location, restaurant, binoculars, field guide author, donut topping (or filling), listing software. Use this information to impress other participants with your close relationship with the leader.

3. Ask the leader to please let you carry his or her spotting scope, explaining that you were thinking of purchasing that very make and model and want to see whether it will be light enough for you. This assures your position at the head of the line (even after you return the scope in five minutes).

4. When birds are seen, impress others with your familiarity with the species by contriving broad particulars. Example: "Wow, I've never seen one so neatly patterned before!" Phrased this way, it hardly matters that you have, in fact, never seen the species before and have no idea what it is.

5. But if in fact you *have* already seen the species, graciously let everyone else see the bird through the scope first. In the event of a life bird, scream (loudly) that it was the species your mother most wanted to see and that her deathbed wish was for you to see it for her. Shout this while you are jumping to the head of the line.

6. When traveling in a fifteen-passenger van, make known your predisposition to motion sickness and confide that you had pepperoni pizza (with anchovies) for breakfast. This will ensure you get the front passenger seat.

7. When traveling in a bus, likewise grab the seat by the door to better manage the pain associated with your recent (choose one: hip, knee, or spine replacement). If birding with a friend, have them take the seat behind you and act as blocker to prevent others from beating you to the door.

8. Carry a bag of salted sunflower seeds (they're cheap). When you see someone else snacking on custom trail mix, cashews, macadamia nuts, dried apricots, Swiss chocolate (and other more expensive items), offer them some

sunflower seeds and, after they decline, graciously accept their reflexive counteroffer. If you bolt the first offered handful down quickly and say, "mmmm," you may get a second.

9. When approached by other participants who, noting your espoused skills, may direct questions to you (since you're monopolizing the leader), gently explain that you, after all, are just a participant too. It's the leader's job to be the leader. Sound deferential.

The Perfect Birder

I guess someone among the R&D folks up in the Creation Department is a reader, or else a fan of mine sent them a copy of my essay "The Perfect Bird." They contacted me, said they liked my suggestions, and did I perhaps have any thoughts re: the Perfect Birder? Well . . .

The Perfect Birder would have 4x vision in front, 2x behind, and night vision and an infrared imaging center in the brain that recreates a full-color image of a bird hiding behind foliage by measuring energy wavelengths.

Heck, the military's been able to do this for years. Think how infrared imaging is going to revolutionize the study of tinamous?

Perfect Birder is able to hear a pin drop on flannel two time zones away, enjoys perfect pitch, has the audio recall of Igor Stravinski, plus automatic playback (for times when some birding companion is talking over the bird), because Perfect Birder can identify the song, call, wingbeat, cough, sniffle, hiccup, sneeze, and belch of every bird on the planet (including those species due to be split in the next two hundred years).

In spring, Perfect Birder's audio control center has an automatic robin filter so that during a Big Day they can hear through the din of the morning chorus.

Perfect Birder is impervious to heat conditions that could leach water out of a stone, cold that would stop molecular motion,

30

altitudes that would rupture the lungs of White-tailed Ptarmigan, and humidity that replicates the atmospheric conditions of the womb.

They have an inner ear that could be put in a blender and maintain its equanimity, a stomach made of cast iron, and an internal stabilizer built into their spine so that no matter how badly the boat is pitching, their binoculars remain rock steady.

Owing to a skin that is part Gore-Tex, part Kevlar, they are impervious to rain, sleet, snow, oobleck, cactus, greenbrier, fire ants, chiggers, ticks. . . . Every hour, tiny pores coat exposed skin with a solution of No. 29 sunblock and ninety-nine percent pure DEET (to keep the black flies from getting in their eyes).

Of course, the Perfect Birder knows every field mark of every bird down to the submolecular level. They have the illustrating skills of David Sibley and enjoy the universal respect of Kenn Kaufman.

Perfect Birder's brother-in-law is Victor Emanuel, who begins every Thanksgiving Dinner conversation with the question, "Well, what tours to obscure, bird-rich corners of the planet should we develop for next year?"

It goes without saying that Perfect Birder is showered by optics from all the top companies (who want their product seen in the hands of Perfect Birder). While Perfect Birder has seen every species of bird on the planet, they never tire of seeing the same birds over and over again (and in the process, keep discovering new species).

Despite their accomplishments, Perfect Birder has the humility of a monk. In the company of other birders, they always let the other person identify the bird first, and in the event of a boo-boo, they cleverly maneuver the erring birder toward the correct identification in a fashion that makes the birder think they corrected themselves. When pointing out a bird to others, their directions are flawless.

On an outing, they always bring enough lunch for two. They always suggest a potty break just when you realize you need one.

They can eat anything, and no matter how many potato chips they consume, they never gain a pound.

They always save the last handful for you.

What sex is the Perfect Birder? That's the best part. They're whatever sex you like.

What could be more perfect than that?

Lawn Flamingo

STATUS: Common and widespread resident species; particularly common where beer is purchased in quart bottles and retired automobile tires find new life as decorative yard fixtures.

DISTRIBUTION: Unlike its tropical cousin, Lawn Flamingo is found almost everywhere in North America, but distribution is both clustered and patchy.

HABITAT: Something of a habitat generalist occurring in a variety of habitat types, but almost exclusively in those altered for human habitation. Places showing high concentrations include driveway entrances and postage stamp–size front yards. Lawns are optional, and in coastal areas this species adapts well to white gravel. Also found in trailer parks and at the entrances to private campgrounds and RV lots. In neighborhoods where streets are designated "Oval" or "Terrace," this species appears most typically in backyards, often in close proximity to in-ground swimming pools, patios, and wet bars.

COHABITANTS: Ceramic deer, gnomes, whirligigs, opulently proportioned posteriors of weeding gardenerlike cutouts, last year's Halloween decorations, plastic flowers, large black dogs chained to porches.

MOVEMENTS/MIGRATION: Nonmigratory permanent resident (except those individuals frequenting RV lots and snowbird parks). In winter, in regions subject to seasonal snow cover, chooses to hibernate beneath a blanket of snow.

DESCRIPTION: A large, long-legged, long-necked, statuesque species (standing much taller than the ceramic gnome, slightly taller than the ceramic Bambi) with a heavy, near boomerang-sized and -shaped bill; small head; long, slender neck (often held in an S-shaped configuration); large, teardrop-shaped body; and extremely long legs. Size varies, but most individuals are considerably shorter than the American Flamingo, from which they may be descended and very closely resemble.

Except (sometimes) for a dark tip to the bill, wholly bright pink. All ages and sexes are similar, except some very old, very faded individuals may be bleached white; albinism is not uncommon. Young, sometimes found in association with adults, are considerably smaller but are otherwise identical to adult birds.

BEHAVIOR: Distinctly sedentary, apparently capable of maintaining the same rigid posture indefinitely. During high winds, sometimes wobbles or sways, but if the bird you are observing takes a step, be assured it is *not* Lawn Flamingo. Usually solitary but occasionally found in small groups (rarely in excess of a dozen birds, except in retail outlets specializing in yard fixtures). In such locations, birds cluster in big pink bouquets, often adorning themselves with oak-tag labels emblazoned with the term SALE.) Young birds, identifiable only by size, most commonly follow adult birds in a single-file line that never moves.

Seems indifferent to humans. Freezes when approached. Has even been observed maintaining statuary demeanor when peed upon by dogs or run over by lawn tractors.

Most commonly stands on flat, dry ground; occasionally on raised flowerbeds and stumps or in shallow standing water. Stance

is most often erect, but some individuals (particularly in regions subject to ground freezes and thaws) display a gravity-defying list that may be maintained for years. Expression: plastic.

FLIGHT: Except on rare occasions when birds are subject to tornadic winds or category-three hurricanes, flightless. During weather-evading flights, birds make spare use of their wings and maneuver by tumbling end over end. Lands clumsily, showing little (if any) regard for habitat selection or self-preservation.

VOCALIZATIONS: Mostly silent. In extremely cold temperatures, sometimes emits a cracking sound (akin to the sound of brittle plastic snapping) and during high winds has been known to whistle, creak, and moan.

PERTINENT PARTICULARS: Locating Lawn Flamingos is often a simple matter of driving down residential streets and keeping your eyes open for yards festooned with yard art. (There seems to be a high correlation between the amount of "art" found in a yard and Lawn Flamingos—the more art, the greater the likelihood that Lawn Flamingo will be present, and the greater likelihood that multiple Lawn Flamingos will be present.)

A dwarf subspecies of Lawn Flamingo, *P. r. wobblensis*, inhabits the dashboards of cars, RVs, and pickup trucks. Regarded by many authorities to be a separate species on the basis of size and mobility (Dashboard Flamingos sway as enthusiastically as Lawn Flamingos do not) the birds are otherwise similar.

Also, while Dashboard Flamingos are mostly solitary, they may be found in association with hula girls and bobblehead dogs, cats, and professional football quarterbacks.

Mine to Give

He was tall and thin and twelvish. He was standing on the bridge over the creek, studiously searching the marshes for whatever birds were to be found. Sitting in a car, reading a novel, was who I guessed was his very supportive mom.

I said, "Hello."

"You're Pete Dunne, aren't you," he said more than inquired.

"That's right," I intoned. "And you are?"

"Adin," he replied. No last name. "I've seen you at the hawk watch at Cape May Point. That's how I recognized you."

"I get there now and again," I admitted. "How about you?"

"As often as I can. Whenever I can get a ride."

I nodded. It's not easy living in a mobile society without ready access to wheels. Still, it seemed that the young birder's ambitions were being subsidized by helpful adults. This, a field guide, and decent binoculars are really all a young person needs to engage a world filled with fun and discovery.

But a quick perusal disclosed that my young friend was one for three in the "need to have" category. The binoculars in his hands were not going to be an ally. He let me peer through them, and the most generous thing I could say was "Mm." As for the field guide in his possession, well, all I can say is that this particular book almost single-handedly builds a case for book burning.

Clearly the parents were supportive, but just as clearly they didn't have much grounding when it came to equipping their aspiring young birder.

Oh so very casually, I made my way over to the car and knocked on the window. "Hi," I said. "I'm guessing that you would be the supportive mom."

She smiled. "He really likes bird watching. Wherever we travel, he brings his binoculars."

"It's a consuming hobby," I agreed, "but listen, the field guide he has is going to hold him back. It's too abridged to be useful, and the layout is miserable. With your permission, I'd like to give him one that works."

"Let me pay for it," she said.

"No," I insisted. "I've got multiple copies and I can only use one at a time. It's mine to give and he'll really benefit from it.

One of the great things about giving is how good it makes you feel. And for about half an hour I was warmed by the thought that that very night there would be this young birder sitting up late, absorbing the wealth of information housed in the book that was my gift.

But the glow of giving began to fade. I found that I couldn't get the fellow's binoculars out of my mind. The best field guide on the planet isn't worth a damn if the image it conveys bears little resemblance to what a birder sees through his or her optics. Without decent binoculars, my gift was compromised.

So the next day when I got to work, I rummaged through my collection of "binoculars sent for testing" by assorted marketing reps. I chose a pair that bore the brand of a company that was supportive of the young birding community, then I scribed a note to Adin, explaining that the binoculars he'd been wishing for had been sent to me by mistake. "Sorry for the mix-up and the delay."

I ran into him the other day. He was wearing the binoculars, and a grin so wide it almost doubled back on itself (but as grins go, it was no bigger than mine).

Why You Are Lucky to Be a Birder

1. You never have to spend money on clothes; you never need to throw clothes away. There is probably nothing that you own (that isn't white and didn't cost more than twenty dollars) that is so out of fashion that it would not constitute acceptable birding attire.

2. When you visit your in-laws, whose television is locked on the Golf Channel, you can sit and feign interest in the match while amusing yourself by keeping a mental list of all the birds heard in the background.

3. If the *New York Times* crossword asks for a seven-letter word (beginning with T) that is the name for the inner flight feathers of a duck's wing, you'll know the answer.

4. When you get stuck in a rush-hour traffic jam on a three-lane interstate and need to know which lane is blocked you can reach under your seat, exhume the spare binoculars that live there, and determine which lane you should be in way ahead of the competition.

5. You get to visit Arizona in August (before the snowbirds arrive and just as post-breeding Mexican hummingbirds are wandering north)—and hotels and RV parks are still offering off-season discounts.

6. You'll never have to decide between two favorite old movies on late-night TV.

7. When you retire, your colleagues will know that what you really want is new binoculars, not a watch.

8. You never have to buy gifts for anyone for the holidays. All you have to do is save all the bird-related junk you get in one season (e.g., bird lamps, bird clocks, bird towels, bird door-mats) and give them away the next.

9. You know why it is that the cardinal (or robin . . . or mockingbird) is throwing itself against your bedroom window and don't have to call the local nature center to solve the mystery.

10. You'll know the names of all the meteorologists on the Weather Channel and can amaze your nonbirding friends (should you have any) with the scope of your intimacy.

11. If you go to a garage sale and find a 1934 Peterson Field Guide, you won't balk at the fifty-cent price tag.

12. You have the perfect excuse to say "no" when your neighbor's cat has kittens and you are invited to take first pick.

13. The prospect of a category-three hurricane striking just west of your coastal town is cause for anticipation instead of anxiety.

14. When you're at a football game and a very attractive (and unaccompanied) fan sits beside you with a pair of cheap new binoculars, you can break the ice by asking whether they know how to use the individual eyepiece adjustment to calibrate the instrument for their eyes.

15. When it's 10:00 A.M. and you are westbound on I-10 just outside Baton Rouge, Louisiana, you'll know not to stop for lunch because you're within range of a bowl of chicken and sausage gumbo at Al-T's in Winnie, Texas—gateway to High Island.

16. When your nonbirding spouse suggests a vacation in Cancun, you can go online and get reservations for (and nonrefundable airline tickets to) Chan Chich Lodge in Belize, explaining later that you must have been confused. Sound contrite.

17. When a cold front hits in late October, you'll be at a hawk watch, blissfully unmindful of all the leaves in the yard that will use the opportunity afforded by the blustery northwest wind to relocate to your neighbor's yard.

18. Season tickets (that get you into High Island, the Beanery, and every National Wildlife Refuge in the U.S.) cost less than one hundred dollars in sum.

19. You never have to worry about losing a ball. You don't have to wear ear protection. You never have to cut bait (unless you go pelagic birding). And the only person keeping score is you.

Effort at Conversation Between Two Birders

SCENE: *The upper bench of the hawk-watch platform at Cape May Point State Park. A middle-aged, middle-weight, middle-income birder wearing a vest that is about ninety-nine percent birding club/festival/destination patches is reviewing his checklist. He is joined by another middle-aged, middle-weight, middle-income birder wearing a cap studded with so many enamel pins it resembles a multicolored hauberk.*

PINS: Mind if I sit down?

PATCHES: Help yourself.

PINS: Pretty fair hawk flight?

PATCHES: Not as good as yesterday. Did you catch any of that? Pretty spectacular, actually. Peregrines galore. Be a shame if you missed it.

PINS [stiffening a bit]: No, I got in last night. But I was here for last Saturday's flight. You must have heard about that. Set a record for peregrines.

PATCHES [stiffening a bit]: No, I was in Duluth. Set a record there, too. Best flight I've seen this side of Corpus.

PINS: Well, those peregrines last Saturday were flying beak to tail—in fact, there was a fifteen-bird pileup over the bunker. Took two hours to clear and the falcon flight had to be rerouted around the lighthouse. Best raptor show I've seen this side of Veracruz.

PATCHES: Talk about congestion, the kettles of Broad-wingeds over Duluth were so thick the National Weather Service mistook them for a line of thunderstorms. They issued a tornado alert. The sky grew so dark that pigeons went to roost. The images of Maurice Broun, Rosalie Edge, and Roger Tory Peterson were seen in the swirling kettles by a dozen witnesses, and Frank Nicoletti ascended to raptor heaven (but he was back in time for supper).

Except for the parry and thrust of imagined swords, there follows a moment of SILENCE that stretches into a period of SILENCE.

PATCHES [pointing to a checklist in Pins' hand]: You a lister?

PINS [looking down at the checklist as if surprised to see it there]: Not really. You see as many birds as I have, you reach the point where it becomes pointless.

PATCHES: I know what you mean. Once you've seen them all [the "like I have" is presumed], you can just concentrate on enjoying them.

PINS [trying to smile but succeeding only in making the corner of his mouth twitch]: My philosophy exactly. I used to try and keep up with the lumps and splits but figured it was pointless. Now I just wait for the AOU to revise their checklist, download it into my computer, figure (give or take a species or three) it's close enough.

PATCHES [trying to smile but succeeding only in exposing his canines]: Plausible idea, but that wouldn't work for me. I've got so many species under review there're records committees in a dozen states working nights. In fact, I think the South Carolina records committee is meeting this week. I hope my warbler sighting in I'on Swamp is on the docket, but, well, that's exactly my point, isn't it?

PINS [not even trying to smile]: Got something of the same problem. There's a bird of mine under review right now, apparently an undescribed flycatcher that distinguishes itself by having no field marks at all. The problem is nobody can decide whether it represents one species or three—and some authorities suggest it may be as many as eight. They'll work it out in time, but until then my list is in limbo.

Except for the grinding of molars, there is a moment of complete SILENCE that becomes a period of SILENCE that becomes an oppressive SILENCE.

PATCHES: Did you hear about the . . . ?

PINS [cutting him off]: Yeah, saw it.

PATCHES: Yesterday it was rediscovered in . . .

PINS: Saw one there before.

PATCHES: Jon and Paul think it might be one of the . . .

PINS: I read the email.

PATCHES: Well, Kenn asked me what I thought and . . .

PINS [interrupting]: Do you know Kenn?

PATCHES: Know him? He named his cat after me.

PINS: Well, I'm godfather to his laptop.

There is a period of SILENCE that defines the difference between mere SILENCE and utter SILENCE.

PATCHES: Well, I've got to go. Packing for a trip to Borneo.

PINS: Yeah, I got to go, too. Finishing up a paper to be presented at the Eilat birding conference.

PATCHES: Well, good birding.

PINS: Yeah, good birding to you too.

END

Letter to the Jerk in the Blue Shirt

This is a letter to the jerk in the blue shirt. I know this description is not very specific. At this point, you still don't realize that this letter is addressed to you. So let me be more precise. This letter is being written for the benefit of the jerk in the blue shirt who was chasing the nesting Black-backed Woodpecker around the tree at Yuba Pass one weekend in June several years ago.

You jerk.

I'm the guy who was standing a couple hundred feet away, suggesting, much too subtly it seems, that you were standing too close to the nest tree and the bird. This letter is intended to express all the things I should have said at the time but didn't.

OK. Let's give you all the benefit of the doubt. Let's say that you are a beginning birder and that you were on a field trip and there was this incredibly unethical field-trip leader who herded a whole bunch of innocent birders beneath the bird's nest tree . . .

And you just lingered, after the group departed, because the bird was so captivating.

Or maybe you just happened to be wandering through that particular aspen grove among all the aspen groves on the planet (the same grove that the birding listserv had posted directions to) and had bumped into the tree and discovered the bird all on your own.

It's possible. Not likely, but possible. However, it really doesn't make any difference.

Why do I say this?

Because when I saw you pursuing the clearly agitated bird around a tree and pointed out "You can see the bird much better from back here," you replied, "But from here I can see every feather."

Realizing that you seemed not to be getting the point, I then pointed out that you were standing right beneath what appeared to be the bird's nest hole. Your response was to point to a hole in an adjacent tree (fifteen feet away), turn your back to me, and bring your binoculars to bear on the still-agitated bird.

Whether you bumped into the tree, were led to the tree, or followed directions to the tree really doesn't matter. The bird was stressed. You were too close. The only ethical course of action was for you to back off and stop crowding the bird. You want to see every feather? Buy a spotting scope. You want to call yourself a birder? Play by the rules.

What rules? Well how about the American Birding Association Code of Ethics? Rule I. BIRDERS MUST ALWAYS ACT IN WAYS THAT DO NOT ENDANGER THE WELFARE OF BIRDS OR OTHER WILDLIFE. Among the five specific no-nos listed under that fundamental axiom is this one: "Keep an appropriate distance from nests and nesting colonies so as not to disturb them (birds) or expose them to danger."

Yes, I realize the term "appropriate distance" is open to interpretation. I'm a writer. I appreciate the subtleties of language. But the adjudicator in this case wasn't you, and it wasn't me—it was the bird you were chasing around the tree, who couldn't leave (because maternal instincts had her tethered to that cavity) and who was telling you by her actions that you were in breach of that most fundamental of birding rules.

Did you tap on that tree to get a look at the bird? I'm not accusing you, I'm asking you. The reason I ask is because my wife is a photographer. And she spent two days trying to get a photo of the bird out of the nest. But being ethical, she didn't crowd the bird's space, didn't resort to any proactive motivation, and didn't grab more than a few quick and distant shots of the bird poking its head out of the hole during one of its infrequent forays.

So how was it that the bird was out of the nest hole so you could chase it around the tree in your effort to see every feather?

Why am I being so hard on you? And what difference can it possibly make whether you (or any other birder, never mind the color shirt) choose to harass some bird? Several reasons.

First, let's talk about stress and nesting birds—particularly rare and aspired-to birds. I personally am of a mind that a little bit of stress in any creature's life doesn't make a whole lot of difference. Stress is natural. It's part of life's package. Birds, like all other creatures, deal with it.

But when you get a situation like we had with regards to a nesting Black-backed Woodpecker, a bird that hundreds of people want to see, the cumulative stress resulting from compounded encounters takes a biological toll on birds, and it puts them, and their genetic dowry, at risk.

Nesting birds try to be circumspect. They try to never draw the watchful eyes of predators. A nesting bird, forced off its nest, behaving with obvious anxiety, is something that no professional crow, raven, raccoon, or Cooper's Hawk is going to fail to note (and capitalize upon)—to the breeding bird's loss and our sorrow.

Second, guilt. Fact is, Mr. Birder in the Blue Shirt, in my nearly forty-seven years of birding I have committed atrocities to birds that I would love to forget or erase. But instead, I will enumerate them here in the hope that they may instill in you an enhanced sense of caution.

At the age of seven, in my persistent efforts to find the nest of a (then) Rufous-sided Towhee, I discovered, much to my horror, that I had found it. I was standing on it.

That's the day I learned that some species of birds nest on the ground.

While leading a field trip in Cape May, I once pished up a young Yellow-rumped Warbler who, fully intent upon the ruckus I'd initiated, suddenly found himself going from 0 to 60 mph in 0.000324 seconds in the talons of a very opportunistic but hardly displeased Sharp-shinned Hawk.

That's the day I learned to refrain from pishing when there are lots of migrating raptors in the vicinity.

And once while searching for prairie birds on an ABA–led field trip in North Dakota, I felt a tug beneath my foot and looked down in dismay to see my boot planted on the outer primaries of a nesting Sharp-tailed Grouse.

That's the day I learned that any interaction with another living thing, even an activity as outwardly benign as birding, courts risk.

But while all of the interactions I've just described were far more injurious to birds than your harassment of a Black-backed Woodpecker, there was a difference. My sad encounters were accidental and based, at least in part, on naïveté and ignorance. Yours was an act of volition (unless you are willing to admit that you were both ignorant and naive—in which case I apologize for calling you a jerk).

And you were one Cooper's Hawk away from having a very sad burden to bear.

Which brings up my final reason for writing this letter: *responsibility.* Our responsibility, as birders, to the hobby and to each other. You see, there are no bird police to walk up and say, "Sir, you are standing too close to that bird's nest tree. Please step back to where that gentleman over there is standing or I'll issue you a summons."

There is no Birding Ethics Court to review the case, hear your side of the story, and dismiss the charges or knock five birds off your life list and sentence you to five weeks of community service building bluebird trails.

There is only us. Us birders. And it is up to us to set and maintain standards for the good of the activity, and for birds. And while I tried to exercise that duty and obligation by urging you to back away from that bird's cavity, I failed in my effort and in my obligation.

That's why I'm writing this now. To state my case, and express to you the things I should have said then and didn't. Because, having made your point and maybe gotten to see every feather, you finally walked away, letting the bird return to the cavity (which she did almost immediately). I figured the issue had resolved itself.

And you know, I would have let the matter drop. I might well have forgotten the whole affair, never given thought to writing this letter. Except that while I was walking away, I passed you going the other way, heading back toward the grove with a friend. You walked by with eyes averted. You were going to harass that bird all over again, weren't you?

You jerk.

Irritation

Don't get me wrong, I love birding. But that is not to say this life-long avocational love affair of mine is flawless or that tiny sand grains of irritation have not on occasion worked their way between the lids of my shell. What irritations? Well, for example:

The type-A birder standing with you on the rickety viewing platform who punctuates short bursts of scanning for (let's use, as an example, a Mongolian Plover) with bouts of pacing—the foot-stomping kind that makes rickety platforms (and tripods) vibrate and distant images dance. Being type A (and impatient), they don't linger; in fact, they leave! Just about the time heat waves are turning the ranks of distant shorebirds into gelatinous mush (making scanning even more futile than it was before).

Boy, is this irritating.

Or how about the person in your birding party who waits until you are about an hour away from your last stop before asking whether anyone else "happened to notice that funny tern flying around"? The one with the red bill that seemed to have white string (or something) trailing behind?

Or the birder who, in response to your question "is anything around?" proceeds to tell you about all the migrants that were here two days ago but "weren't seen yesterday." Or the native resident offering directions to the migrant trap where all the birds were seen (two days ago), whose reference points are so esoteric even a local birder would be confounded.

"OK, you go to the bridge where the Cave Swallow attempted to nest with the Cliffs eight years ago. Know where that is?"

"No, I don't."

"That's OK. Then you take the turn to the reservoir where the Yellow-billed Loon was reported but not verified in 1965. You must know where that is."

"No. I live two time zones from here."

"Oh."

But some of my irritations are much closer to home. In fact (I'm embarrassed to admit) some are even part of my home life. I don't know what I did to bring this on but wife Linda has recently taken to faking birds. For no apparent reason she'll stop, point, and say "hear that?" And I'll say "hear what?" Then she'll make up some bird name like "Blackpoll Warbler" or "Ruby-crowned Kinglet." You can imagine how irritating this is.

It's almost as irritating as having some seven-year-old step onto the hawk-watch platform, look up into a pure blue sky, and ask "what's that one?" Bringing your binoculars to bear, you find an American Kestrel in orbit around the international space station. "That's a kestrel," you say, and the kid screws up his face, shakes his head, and says, "No, not the kestrel, the big hawk above it."

And there's nothing there.

Sometimes I think that people who profess an interest in facilitating my birding are secretly working against me. Who was it that invented the one-way refuge auto-tour route? And why is it that mega-rarities always turn up at the midpoint right after you've driven past?

And what genius decided that good birding vantage points should be blocked by elaborate shacks with narrow viewing slits? Structures that are appropriately called bird "blinds" because that is the way you feel when you're sitting in one. Hello, you parks people, you birding trail designers, you people with grant money to burn, did it ever occur to you that birds simply get used to a pattern of human behavior and so long as everyone stays on the trail or the open viewing platform, birds will just go about their business? Yes, I know blinds (or "hides") are popular in the UK, but maybe their popularity there has less to do with any presumed sensitivity on the part of birds than it does with meeting the comfort needs of people birding in a cold, rainy climate.

And I know that this a petty thing, but more and more I find myself irritated by birders who link themselves and birds with possessive verbs. Example: "I still need Bohemian Waxwing." Or: "I got the rail." I am particularly irritated when I do it myself and . . .

. . . when I check into a hotel that offers a pool, sauna, wet bar, exercise room, free internet access—but not the Weather Channel.

And finally (and I know this is really petty), why is it that birders habitually direct other birders to birds in flight by telling them that the bird is moving "left to right" or "right to left"?

If a bird is heading left, it stands to reason that it is coming from the right.

Right?

Confessions of a Listing Heretic

Birding (like civilization) is founded upon an adherence to common standards and practices. It's what distinguishes us from the savages, from people who cheat at solitaire. It's what makes Life Lists, Big Day scores, and Christmas Bird Counts possible, comparable, year to year, team to team, person to person.

What would we have if some people went around spuriously counting Red-legged Black Ducks as full species and other people just up and decided that Black Ducks and Mallards were conspecific?

Anarchy. That's what.

Would you want to bring up a young birder in a world where some people counted the Green Kingfishers perched on the Mexican side of the Rio Grande?

You would not!

On the other hand, there's a lot of gray out there between the black and white. Some of the rules governing what and how birds are counted are hard and fast. Others are open to interpretation. And sometimes even the most conscientious birders have been known to hedge just a wee bit in the name of a good bird or a good cause. In fact, some of my favorite birders have advocated positions with regards to the legitimacy of certain birds that are downright heretical.

55

Take, for example, Dr. Ernest Choate, who was for many years the keeper of the Cape May Bird List. For years, Ernie carried a Manx Shearwater on the list even though the bird was seen off the Lewes, Delaware, breakwater—twelve miles from Cape May. His rationale: "We can count anything on the Cape May list that can be seen from shore. On a clear day, we can see Lewes. Therefore . . ."

This same brand of logic has led a friend of mine to practice a most convincing reinterpretation of the rules governing Christmas Bird Count circles. The northern limit of the Cape May Count passes through Goshen. From Goshen, an observer can see Jakes Landing, three miles away, and if you are gifted with good eyes and superior skills, you can even see and identify hunting raptors at that distance—and my hawk-watching friend can and does. But, for the sake of accuracy, and in disregard of the three miles separating Goshen from Jakes Landing, my friend makes it a point to go over and "ground truth" his observations.

Maybe it's the spirit of the season, but Christmas Bird Counts tend to bring out the loose interpretationist in us all. Nobody will ever know how many count "circles" are dotted with "pimples," little bulges in those pure-hearted lines that sneak into key birding locations. Why the venerable Barnegat CBC has half of Long Beach Island dangling off its southern rim—a veritable geographic hemorrhoid! And in the best tradition of Christmas Counts, some of the birds that get reported at the roundups would prompt Diogenes to blow his brains out.

I'm a practicing and self-avowed listing heretic myself, and this is what gives me standing to throw stones. Once, on the Wallpack Valley Christmas Count, I went out for a bit of predawn owling. A heavy snow had fallen overnight, stopping just after 2:00 A.M. While I was driving along (of course in the count circle!), my headlights picked up something lying on the road. Closer examination disclosed a dead screech-owl. The bird was lying fully atop the snow. Therefore it had been killed *after* 2:00 A.M. And therefore it had been alive and in the count circle on the count day.

Screech-owl. Check.

And then there was the Black Vulture claimed by the Guerrilla Birding Team in the 1988 World Series of Birding. The bird was breeding in a shallow cave, well off our route. A midnight visit was the only alternative, but there didn't seem to be any way to approach the cave without flushing any adult within—something we patently refused to do. Midnight found me fifty feet from the cave entrance, playing a flashlight in the trees above the cave, hoping to find one of the pair at roost. But it wasn't to be. Just as I was preparing to leave, a *hssssss* emanated from the cave entrance. I left. With a smile.

Black Vulture by call. Check.

This same spirit of concern is what led ultimately to a World Series of Birding rule change regarding nesting raptors. A very accessible goshawk nest had become known to many of the event's participants, and concern for the bird's welfare mounted. To save the bird from serial harassment a ruling was made. Any team that knew the location of the nest need only drive to the grove, park, wait five minutes, then drive away. The bird would count sight unseen. How's that for creative counting?

Ten Birds for Walter

I must have passed that little pioneer church and its cemetery thirty times and never noticed it. Funny how being on a bicycle makes the world more engaging.

It was early autumn and midafternoon. In Paso Robles, heart of wine country in California's Central Coast, this means hot, dry, and pretty birdless. That's why I was biking, not birding.

I parked the borrowed three-speed. Mastered the church fence latch. Made my way past the planted junipers and toward the array of limestone markers.

Some people are troubled by cemeteries. Me? I find them tranquil and inviting. A place for contemplation. A window to the past (if not to another world).

And while most birders recognize the birding potential of cemeteries, they are as a rule not as crowded as most of the well-known birding hot spots (Mount Auburn Cemetery excepted). As such, they might hardly be considered a place to meet new birding friends. But this cemetery and this time proved different.

The marker that caught my eye was one of three. They bore testimony to Otis A., Evan P., and Walter M. Stoval—sons of F. M. and M. C. Stoval, whose graves were only an American Robin's strut away. As the inscriptions on the stones attested, the three Stoval boys, ages ten, eight, and four, had all died within weeks of each other. The oldest, Walter, died on May 28, 1885.

At this point you must be wondering why my impromptu communion with Walter Stoval merits mention in a collection of birdwatching essays. The reason was chiseled in stone. It read:

Jesus had gently called him away.
Dear little Walter meekly obeyed.
Bidding farewell to the birds and his friends.
Enduring patiently onto the end.

The birds and his friends! This ten-year-old kid from a bygone age was a bird watcher!

Make that "bird lover" or "student of nature." The term "bird watcher" didn't appear in the lexicon until the early middle of the twentieth century.

Of course I felt a sense of kinship. Ten was about the time I got really jazzed about birds myself (even though I'd started keeping notes when I was seven, even though the term "jazzed" didn't exist when Walter developed his bond with birds).

That his interest was intense was manifest. Even today, in an age where bird watching is socially acceptable, how many deceased birders do you know who have joined their names and their avocation in stone?

Here lies John Doe, Birder

Mary Contrary, 1954–2007,
North American Life List: 712 (NIB)
NIB = no introduced bird species

Of course I felt a little sad, too. I mean, here was a kid who was watching birds back when Spencer Baird and Robert Ridgway walked the earth. Had he lived, he would have been a contemporary of James G. Cooper, the noted California ornithologist.

He probably had California Condor for a yard bird. Had he pursued his passion, he might have discovered the nesting secret of

Marbled Murrelet decades ahead of schedule, or come up with the idea of a field guide before Roger Tory Peterson was even born.

We'll never know. All that potential was cut short. All the hundreds and thousands of hours of fun and discovery that might never have been realized. All . . .

"Right," I said to the stone. "Let's see if we can make some amends here.

"Walter, it just so happens I'm a birder visiting from New Jersey and I'd love to do a little birding with a local. Mind if I join you for an hour or so?"

Walter didn't seem to mind at all.

"Tell you what," I said, easing myself into a sitting position next to Paso Robles' answer to Alexander Wilson and Gilbert White, "let's make a game of it. See if we can tally ten species on this hot afternoon. One for every one of your birthdays. See if we can get them in an hour or less. I'll scan; you keep score. Do we have a team?"

No objections from Walter. The match seemed made in heaven.

Ten species in the middle of Wine Country on a hot afternoon was going to be a tall order, but I was pretty sure the kid and I were up to it. The first bird was a shoo-in.

"Yellow-rumped Warbler," I said, pointing into the cedars behind us. "Got it by call."

Nothing from Walter. No kudos. Not an attaboy. Then it hit me.

"Right. Audubon's Warbler to you. Lumped 'em back in . . . or from your perspective, not until 1972."

The second bird was European Starling, and of course this took a bit of explanation.

"It goes like this. There was this guy who thought it would be really fine if all the birds mentioned in Shakespeare were introduced to North America. Unfortunately, he succeeded with this one."

Nothing again. Maybe Walter was just hard to impress. Maybe . . .

"Of course! That's a Life Bird for you, isn't it? Sorry. That puts a different spin on things.

"Congratulations," I amended. Probably the first time anybody has offered congratulations about the sighting of a starling in a hundred years.

Number three was a Red-tailed Hawk. A nice western adult. California's Central Coast is Red-tailed rich, and a birder would have to be dead and buried not to be able to find one in five minutes or less. Walter (and I) managed in just under four. Speaking of which . . .

Number four was Golden Eagle, and it was a little tougher than the Red-tailed. In fact it was *w a a a a y* up there. Binocular range only.

"Unless you've got really good eyes, Walter, you're going to have to accept my word for this one." On the other hand, given the bird's altitude and Walter's perspective, it might be that he was enjoying a pretty good view of the bird (albeit a dorsal view).

Turkey Vulture was number five. House Finch an easy six.

"If you'd been born in the East, you would have been an old man before you caught up to one of these things. Lucky for you the birds are California natives."

It was getting tougher now. Four birds and thirty minutes to go. It took ten minutes before a Northern Flicker, number seven, looped across the horizon.

"Good one, Walter. Thematically apt. Some day that's longer back for me than it is ahead for you, some lad about your age named Roger Peterson is going to run into one of these woodpeckers at a place called Swede Hill, and the encounter is going to change the world."

"Course," I added, "you might have beaten Roger to the encounter, and the punch. And then there would have been a Western guide to birds before an Eastern guide."

About this time, two blackbirds offered a momentary view before disappearing behind the trees—and while they were probably Brewer's, I couldn't be certain.

"Want to count 'em?" I asked.

The silence spoke volumes.

"OK," I agreed. "But you're not making this easy."

Actually the next bird, Mourning Dove, was a very easy number eight. Given the habitat, I couldn't understand why such a common and thematically apt bird hadn't made the list before this.

The adult male Rufous Hummingbird that shot past was an unexpected bonus and number nine. I looked at my watch and noted, with growing anxiety, that we had only five minutes left.

"You can still change your mind about the blackbirds," I told him.

Silence. I guess he was thinking about it.

With two minutes to spare I heard a warbler *chip* note overhead that recalled Black-throated Green Warbler, but, of course, given our location, wasn't. Binoculars disclosed a . . . "Townsend's Warbler!" I shouted.

Walter was speechless. Not that I blame him. Back when I was ten, identifying fall warblers in flight was just about inconceivable. Now . . .

"Nothing to it," I said to Walter. "Nothing a couple hundred years of bird-watching tradition can't handle. Heck, I know nine-year-olds who can do it."

It was getting late. I'd promised Linda I'd be back by four.

"Listen, Walter," I said, "it's been great birding with you." I tried to think of something to add, but none of the standard lines ("take care," "happy trails") seemed to fit.

"Be thinking about you," I said and took my leave, and I did and I do . . .

Think about Walter Stoval. Bird lover. Age ten. Who resides in the Wine Country of Central California. Who might have changed the face of bird study. Who just got starling as a Life Bird. And whom I go birding with now . . . and maybe again.

Aspiring to Martins, Settling for Wren

I walked up to the counter and smiled.

The man behind the counter returned the smile.

I said, "I'm thinking of putting up a birdhouse; attract some birds to my yard."

He said it sounded like a fine idea. Did I have any particular species in mind?

I said, "How about Purple Martins?" That's when things got complicated.

"OK," he said, "I need to see two forms of ID—one with a recent photo."

"I plan to use cash," I told him.

"It isn't a matter of credit," he explained. "It's a matter of responsibility. I need ID to do a background check."

"To do *what*?"

"To conduct a background check," he repeated. "Putting up a Purple Martin apartment is a serious matter. Not all people are emotionally and psychologically equipped to assume the duties and obligations associated with becoming a Purple Martin landlord."

"OK," I said, digging into my billfold. "But isn't this sort of extreme?"

"Sir," he said, "I should advise you that my discretion in this matter is absolute and that anything said in my presence may affect my decision regarding your eligibility to purchase a Purple Martin apartment. Don't you realize that over most of North America Purple Martins are wholly dependent upon humans to provide suitable housing? Their survival is contingent upon their partnership with our species."

"Wouldn't an alliance with woodpeckers serve them better?" I suggested.

He didn't smile. What he did was hand me a questionnaire and an officious-looking document entitled "Covenant to Assume All Obligations, Duties, and Responsibilities Concomitant with Purple Martin Landlordship." Then he left to run the background check.

The questionnaire wanted to know:

Whether I was now or had ever been censured for negligence relating to the maintenance of Purple Martin colonies in this or another state;

Whether I was now or had ever been a member of the House Sparrow Alliance or Starling Firsters, or any other organization whose objectives, stated or tacit, might undermine the martin social order;

Whether I was phobic about feathers, bird droppings, or feather lice;

Whether I had any injuries or physical limitations that would prevent me from performing the duties incumbent upon Purple Martin landlords. These duties, summarized in an accompanying "Covenant to Assume," included my promise:

To erect the box in accordance with mandated guidelines;

To lower the box weekly to remove insurgent pest species;

To replace nesting material halfway through the nest cycle;

To measure nestlings and plot their growth and development . . .

"It must be signed in purple ink in the presence of a witness," the returning clerk advised.

"The background check came up negative. Could I please see the birthmark on your left buttock?"

"Right buttock," I said. "No, you can't."

"Just testing," he said, reaching into a drawer, removing a set of cards, holding the first in front of my face.

"Look at the inkblot and tell me what you see."

"Mother Teresa dropping mealworms into the open mouths of nestling martins?"

"Good," he said. "How about this one?"

"Lenin removing the capitalist yoke from the necks of down-trodden Purple Martin masses?"

"Excellent!"

After passing this, and a stress test, and successfully field stripping and reassembling a Purple Martin apartment while blindfolded, I was allowed to take the Pledge of Martin Management . . .

"On my honor."

"On my honor."

"I will do my best."

"I will do my best . . ."

Then I was ushered to a display area and introduced to a basic five-star-rated Purple Martin apartment complex.

"How much?" I asked.

"Four hundred dollars."

"Four hundred dollars?!!!"

"That includes the pole, pulley, and predator guard. Installation is extra."

"Isn't there a cheaper and less-consuming way to attract birds to my yard?"

"Sure," he said. Walking over to a display. Coming back with a basic-looking bird box. Handing it to me.

"What's this?"

"Wren box."

"Do I need to take a course in environmental ethics or submit to weekly inspections by the Wren Police?"

"Nope," he said. "Just hang it up. The wrens do the rest."

"Sounds like Purple Martins could learn a lot from wrens," I joked.

The clerk didn't smile.

"Sir, let me remind you again that anything said in my presence that reflects negatively upon your desire or resolve to assume . . ."

I bought the wren box.

Talkin' Trash (Birds)

"**O**h," (name withheld) said, rolling his eyes, "where I come from, that's a *trash bird*."

Now I know what he meant. And chances are you know what he meant. But that doesn't mean the expression "trash bird" is not free of interpretation. For instance . . .

This past winter I joined a bunch of Delaware Valley Ornithological Club members on their annual gull outing. To the Jersey Shore (some might assume) or Niagara Falls (perhaps) or some nice strategic hydroelectric dam. No, actually, it was on the banks of the Delaware River across from an open landfill so spacious it spans zip codes and so towering it merits contour lines on a geologic survey map.

And there were gulls. Tens upon tens of thousands of gulls—a veritable gull maven's paradise. Notables included half a hundred Lesser Black-backeds, a dozen Icelands, three Glaucous (including one stunning adult), and at least one very defensible Thayer's Gull.

Every one of them a certifiable trash bird—if trash bird and garbage bird are allowed to be synonymous (and since we are talking about a group of birds that can eat darn near anything, they might as well be synonymous).

Open landfills are much the source of another apt application of the term "trash bird," this one relating not to orientation but to adornment. If you've spent any time watching gulls (or cormorants, or loons, or waterfowl) you have no doubt had the sad experience of seeing a bird with a six-pack holder or some other unwanted item around its neck. My life Glaucous Gull was wearing a sardine can on one foot (made identification very easy). On another occasion an urgent plea for aid from a local golf-course groundskeeper, re: a Canada Goose and an offending six-pack holder, led not only to the "trash bird" but also to a Greater White-fronted Goose (Greenland race), which is the very antithesis of a "trash bird" in northern New Jersey.

Said the greens manager: "He took one look at this weird goose in with the flock and forgot all about the poor . . ." Actually, I didn't forget the goose. But I did quickly assess that insofar as the garnished bird was alert, healthy, and capable of flight, my chances of removing the offending plastic were nil.

And while you and I (as caring individuals) might wish that birds and trash never mix, a whole host of birds clearly have other ideas. It is a rare robin's nest that doesn't have some bit of discarded paper tucked in with the adobe, and Great-crested Flycatchers are as like as not to substitute shed plastic for shed snakeskins. I've seen some Osprey nests that could have served as the screen set for the old TV sitcom *Sanford and Son*, nests whose collection of junk included beach-chair webbing, plastic shovels, shotgun shells, and fishnets—all manner of things our species considered trash but that Ospreys considered decorative treasure.

You think this is deplorable, don't you? I quite agree; in fact, it's even worse. On occasions the situation is tragic. There are documented cases of young Ospreys ensnared in the fishing line that the parents brought into the nest.

But consider the male Satin Bowerbird, a blue-eyed, crow-sized species that advertises his worthiness by first building an elaborate tepee (or bower) and then by decorating the interior with artifacts that capture the color of his eye in the hopes of catching hers.

I've seen just one Satin Bowerbird's nest. It was a dandy. But what I remember most was not the structure but the treasures within. There, nicely arrayed within the bower, was a seductive stash whose principle treasure was a collection of blue tops from old Bic pens (whose color matched the eyes of the Satin Bowerbird perfectly).

Very possibly the tops were lost or discarded. But imagine the consternation of some poor lodge employee who could not figure out why pen tops kept disappearing from her desk.

And last but certainly not least are the "trash birds" that are wholly the former and none of the latter. I'm talking about all those tantalizing objects that have none of the earmarks of birds but under the catalytic influence of distance, heat waves, and wishful thinking assume avian form.

No one will ever know how many Clorox bottlers have been transformed into Snowy Owls on how many beaches. I've done it a dozen times, but I have also figured out how to tell the difference between a real owl and a pseudo-owl.

What I do is search for the nearest Herring Gull and gauge how close it is standing to my presumed owl.

If the gull is more than a hundred yards away, it might well be an owl. But if the gull is in close proximity, chances are my "owl" is just another trash bird.

Change, Change, Change

I passed a milestone this morning. Literally. I also came to a level of accommodation that it takes, it seems, a lifetime to acquire.

Motoring along Turkey Point Road, Cumberland County, New Jersey, I drove past a Bald Eagle. It was perched about a hundred feet off the road, magnificent as could be.

"But what," you might be thinking, "is so momentous about a Bald Eagle?"

Yes. Precisely. Thank you for making my point. What made this encounter notable was not the presence of the bird but my reaction. As stated, I drove right by it (as I might any other common species I didn't wish to disturb).

Forty years ago this would hardly have been the case. Forty years ago, when Bald Eagles were downright rare across most of North America, I would have stood on the brakes. Slammed binoculars to my eyes. And reveled at what would have been my most intimate look at a Bald Eagle.

But in 2014 A.D., and through sixty-three-year-old eyes that had, in the last hour alone, observed six individual eagles, I gave the perched bird no more than an admiring glance and drove on.

Had it been an American Kestrel or Rough-legged Hawk, you bet I would have stopped! Wintering kestrels in these parts are as rare now as Bald Eagles were back when. As for Rough-legged Hawks . . .

71

Haven't seen one on the bayshore in three, four years. Used to have them every winter though.

This essay is about change. I've seen a lot of it. Seen a world go from a time when Golden-winged Warblers were as regular as summer to a time when they are as common as snow in July. Seen New Jersey's woodlands go from zero Wild Turkeys to encounters, such as the one I had recently, where I counted 244 birds in a single field.

Locked in my memory is a time when coveys of bobwhite infested every South Jersey farm and Ruffed Grouse haunted every North Jersey wild-grape tangle.

In winter, Evening Grosbeaks were so common they'd break a Wall Street broker's birdseed budget. And in my youth, Tree Sparrows flocked to my feeders.

Last year, bobwhite was removed from the New Jersey state game list. They have all but disappeared in the state.

Grouse? Most teams miss them in the World Series of Birding now. I haven't seen one in over a decade.

As for Evening Grosbeak, the last time I saw one at a feeder, Linda and I were spending our first winter together. Last July, we celebrated our twenty-fifth anniversary.

But yesterday she ran upstairs to announce that there was a Tree Sparrow at the feeder. It had snowed overnight (a momentous occasion in itself).

Now the big question. Why have all these species declined?

Easy answer. Because things have changed.

What things? Lots of things.

Perhaps whatever supported or motivated these once-common species isn't supporting or motivating them anymore. Or maybe the habitat that sustained them is for some reason no longer sustaining. Or perhaps their population is doing poorly—maybe because the population of some other species, either a competitor or predator, is doing well.

The natural world is ever changing. The elements change with it. It is only memory that is stagnant, and, perhaps, our desire to have the world always as we remember it.

Take Geese

I remember the first pair of Canada Geese that settled into the ponds behind my parent's house in northern New Jersey. It was April 1962. The winter ice had just melted. I couldn't believe my fortune, and of course I assumed proprietary ownership over the birds.

My ponds. My geese.

I wrote to the governor to make him mindful of this momentous event. I urged him to cancel the goose-hunting season in New Jersey until the birds could become established.

I guess he didn't get the letter. Hunting season went on as scheduled.

But, as it turned out, the geese did just fine without the moratorium. In fact, in a remarkably short time, New Jersey's Canada Goose population reached plague proportions.

In 1978 I took part in an annual affair at Brigantine National Wildlife Refuge called the "Goose Roundup." When geese were molting and flightless, scores of volunteers would form a skirmish line and march across the impoundments.

"March" really means "wade," up to our waists, and part of the fun was that everyone got pretty mucky.

The geese were herded into traps and then shipped to whatever National Wildlife Refuge would take them. The roundup was ultimately canceled, not because it wasn't successful but because the geese were. By the early 1980s there wasn't a refuge to be found that wanted Brigantine's geese. Everyone had plenty of their own.

The Ups and Downs of Raptors

Canada Geese were not the only species whose welfare concerned me. Like most environmentalists in the sixties and seventies, I was close to militant about restoring raptor populations depressed by the widespread use of DDT.

I wrote letters. Joined organizations dedicated to the protection of raptors. And committed hundreds of hours to counting migrating birds of prey, believing, fervently, that these monitoring efforts would benefit raptor populations.

Which in fact they did. The data from the hawk count in which I participated led to the acquisition of the South Cape May Meadows by the Nature Conservancy and was also used to justify the acquisition of Higbee Beach and Hidden Valley Ranch.

These protected habitats, all in Cape May, are used by millions of migrating birds (and thousands of birders) today.

In Cape May, in the seventies and early eighties, raptor totals were high, averaging more than sixty thousand birds of prey per fall. Then in the late eighties and nineties things began to change. While the totals for some species, most notably Bald Eagles and Cooper's Hawks, continued to climb, other totals took a different tack.

Peregrine Falcon and Osprey numbers leveled off. Kestrels began a decades-long decline. And Sharp-shinned Hawks? Cape May's bread-and-butter bird?

Their numbers fell. Then rose again. Then fell again. And while Sharp-shinneds continue to be Cape May's most abundant raptor migrant, the present autumn totals of approximately fifteen thousand are only a quarter of what were passing during those early years, the years during which Sharp-shinned numbers were rebounding after DDT was banned.

It makes perfect sense when you think about it. Have a problem. Identify, then fix the problem. Board resets itself.

And Now for Something Really Rare

After driving past the perched Bald Eagle, I went off in search of a wintering kestrel I'd located several weeks earlier. That's how I found the flock of turkeys. They were in a field with twenty Eastern Bluebirds, another species that was uncommon in my youth but seems to be doing fine now.

Unlike American Kestrel.

Am I concerned about the kestrel? No. I'm sorry that one of my favorite birds is no longer common. But the universe seems to move in ways that are indifferent to my likes and dislikes. I am no more "concerned" about kestrels that are declining than I am "concerned" about ravens, whose numbers are increasing steadily in New Jersey.

What I am, more and more, is fascinated by this dynamic, ever-changing natural world even when I do not understand the dynamics. If I have any concern, it is that my species, as much as possible, maintain a level playing field on which the players in the natural world can strike a natural balance.

Many years ago, while working on a book entitled *The Feather Quest*, I dedicated the penultimate chapter to a discussion of birding's future, with an underlying focus on the human impact upon bird populations.

The setting was Roger Tory Peterson's study. The literary carriage a conversation with birding's Grand Master, who as an octogenarian had a great deal of perspective about birds and bird populations. Said the man who had dedicated his life to the welfare of birds, "You know, I saw the first Forster's Terns return to New York City."

After the destructive practices of the millinery trade were halted, Forster's Terns, as well as many other persecuted species, repopulated their historic ranges. It took several decades, but Roger and the members of his generation were there to witness this natural rebound, just as the members of my generation witnessed the recovery of raptor populations after DDT was banned. This recovery, too, took decades.

I never did find the kestrel. Sad. But then, four decades ago, I would not have seen numbers of eagles and all those turkeys.

What a dynamic world we live in. And fascinating.

First Walk

I don't get many young birders on my weekly walks. Maybe it's because the walks fall on Mondays and school is in session. Maybe it's tough getting nonbirding moms or dads to agree to the requisite transportation. Or maybe things haven't changed much since I was a kid. Back when my life was governed by things like homework, bird watching wasn't considered cool. In fact, being a bird watcher ranked right up there with liking girls as a way to get yourself blacklisted with the in-crowd.

So the young eight- or nine-year-old birder among the ranks of adults one Monday last June would have caught my eye anyway (even if he hadn't stuck to my side like a burr on corduroy).

He distinguished himself in other ways, too. Like the way his eyes glided over the habitat, searching for birds. Like the speed with which he brought his binoculars to bear. Like the casual way he pinned names to fast-moving forms.

Clearly the kid took his birding seriously, and just as clearly relished his proximity to the guy at the head of the line, me. And then . . . I remembered . . . my own first time at the head of a line.

I was in fourth grade. That would have made me about nine. It was spring, probably early May, and our teacher was just as spring-bitten as the rest of us. When she suggested that we ignore the lesson plan and go for a walk around the school's wooded perimeter, my classmates and I ran for the door like some multilegged beast.

We'd gone less than a football field's length when the loud, persistent singing of a curve-billed, spot-breasted bird caught the ears, then the eyes, of my teacher.

"Does anyone know what kind of bird this is?" she asked.

"It's a Brown Thrasher," I said, before realizing the consequences of such a disclosure. Everyone gasped.

"Was that you, Peter?" Mrs. Manning asked, and not without reason. I wasn't exactly a standout student. In fact, I was one of those shy quiet types who slide low in their chairs when teachers ask, "Who knows how many times five goes into . . ." You know, the kid always picked last when they divvy up sides for kickball. That was me.

"Yes," I admitted.

"Do you know any other birds?"

"Yes," I said again.

"Would you like to show us some?"

"OK," I said, because I liked seeing birds more than anything, and because in my whole life nobody had ever asked me to lead anything before.

And so there I was. At the head of a line of twenty-five students. Except pretty soon it stopped being a line and classmates were crowding around me asking "What's that one? . . . what's that one? . . ."

They were just the common birds around our school. But I knew them, and my classmates were impressed.

"Would you like to lead for a while?" I now asked the young burr of a birder at my side.

"OK," he said, and he did. Took a two-step lead. Pointed out a Cedar Waxwing perched on its namesake tree. Drew everyone's attention to an Osprey circling overhead.

"Let me know when you want to take over again," he said, one leader to another.

"OK," I said. Boy it's great being at the head of the line. Way better than getting picked first for kickball.

How Waterfowl Got Their Colors

If I were one of those wise old Native Americans, I'd craft a legend explaining how ducks came to don so many fantastic colors. I would explain how ducks were made early in the creation process (before the folks in R&D had solved that problem relating to bills and getting them pointy, and before the design team had given much thought to using colors). While not much to look at, these proto-ducks were gifted with beautiful voices, could sing rings around thrushes. All the songbirds were understandably jealous.

They petitioned the Great Spirit to order a recall (which, wanting to keep peace, he did). So he retrofitted ducks with voices exhumed from the recycle bin, but then, feeling guilty, he contacted the R&D team and asked whether they had anything to offer ducks as compensation.

It so happened that the team was wrapping up work on the "Birds of Paradise" series and starting a top-secret project dealing with iridescence (code name STARLING). They said inventory was tight, but if the Great Spirit could limit his paint job to a dab here and there, and maybe treat just one sex, they might be able to make their stock stretch.

First bird in line was the drake Mallard (who wasn't a bad-looking bird as it was). He studied the options and asked to have a green head.

It glistened. It glowed. If you painted your house a cone cell–jarring green like that your neighbors would sue. But on the Mallard it was stunning.

Other ducks thought so, too. Drake shovelers, Greater Scaup, Common Goldeneye, and Common Merganser all opted for green heads.

American Wigeon and Green-winged Teal got to the head of the line just as stocks of green were running low. They were limited to green eye stripes (instead of getting all-green heads), but the teal (being a tough negotiator) pointed out that because of his smaller size he was getting less paint than the larger ducks.

That's how the teal got a green speculum, too. He was such a good negotiator he managed to secure a green speculum for the female as well.

Ring-necked Duck, Lesser Scaup, and Barrow's Goldeneye all chose to make a fashion statement with purple. This time, Blue-winged Teal was last, and once again the stock ran low. They had to thin the mix. That's why the head of drake Blue-winged Teal shines both purple and rose (the rose is the undercoat showing through).

Rusty red was the next craze. It made a splash with Canvasback, Red-head, and especially Cinnamon Teal and Ruddy Duck (they wanted their whole bodies done).

The Green-winged Teal (the tough negotiator) got back in line and demanded that his head be painted red where it wasn't already green. He got his way (after agreeing to leave the female out of the equation this time). When the wigeon saw this, some went back and demanded to have their whole heads painted red. The Great Spirit was afraid it would start a trend, but he relented (with the result that some wigeon became Eurasian) and sure enough, wouldn't you know, shoveler got back in line to change his green head to red, too. This time the Great Sprit said no (but they had lots of red, so they painted a red swath down the male shoveler's side— just to keep him happy).

Wood Duck and Harlequin Duck? The plans were submitted by a couple of school groups as part of an interscholastic science competition. Nobody thought they'd actually go into production.

While this may (or may not) be the way ducks got their colors, one thing is certain. For sheer WOW, nothing beats the colors in the plumage of waterfowl. The songbirds might have had their way. But it's every birder armed with binoculars and a spotting scope who is the real winner.

Revealed at Last: Why Blue Jays are Blue

The snow, which started falling about first light, ended shortly before noon. Sitting in my tree stand, cloaked in four inches of fresh powder, I was as close to being one with the surrounding North Jersey woodlands as a person can hope to be.

It was then that I saw the jay. Two hundred yards away. About twenty feet up in a tree. The only splash of color in a woodland gone black and white, cryptic as a strobe on a moonless night.

"How," I marveled, "can an electric-blue bird get away with such telltale plumage in a landscape filled with hunting eyes?"

Or put another way: What possible advantage can there be to being so superconspicuous outside the breeding season?

And I believe that it was the simple formulation of these questions (and the time to ponder them) that led to the answer. In the process I also garnered a new respect for a bird I've watched all my life but only thought I knew.

Blue Jay 101

As most readers know, birds come in a variety of colors and patterns. Some are showy, some cryptic, and not a few are both. A number of species sport an eye-catching array of feathers during courtship (because they want to be seen and they want to impress). But breeding season over, they shed their courtin' clothes

ASAP, slipping into something less flashy. In a world filled with hunting eyes, standing out is a liability.

And then . . . there is the Blue Jay. Male or female, young or old, the birds are just about the biggest, brightest, most flamboyant, most conspicuous birds in the eastern forest. A visual treat for birders. A liability waiting to happen in a winter landscape.

It didn't make sense. It seemed to go against logic, reason, not to mention natural selection.

But just as a creature's greatest strength is also its greatest point of vulnerability, a creature's presumed liability can also serve as a strategic asset.

Time on My Side

The great thing about sitting in a tree stand for a week is the latitude it gives you to study the world around you. When I'm birding, I'm more or less shopping. Walking down the aisles. Picking out the species. Pinning names to them. Putting them in the shopping cart of my mind. But when I'm sitting in a tree stand, it's like sitting down at the dinner table—a natural banquet. There is time to savor. Time to get your fill. Time to study the antics and behavior of . . . chickadees, titmice, creepers, nuthatches, woodpeckers, and . . .

Jays. It's a poor woodland that doesn't host a troop of jays in winter, and my favorite woodland in Hunterdon County, New Jersey, is no exception.

There were five jays in all. Age, sex, genetic links— indeterminable. But I'm going to call them a "family group." In late summer, after fledgling, I've watched certifiable family groups of Blue Jays foraging through the forest. I speculate that these groups remain intact throughout the winter.

Because it makes sense; because, as I now interpret it, staying together holds a strategic advantage for jays.

So for several days one December I got to watch this family group of jays moving through my woodlands. In the morning they'd head east; in the evening, what I presume to be the same five jays would move through heading west—foraging as they went, coming and going.

It was obvious they had a pattern, a routine. But it wasn't until day four that I realized that there was more to it than that. There was also order, discipline. Their movements were not random; they were reasoned, patterned.

The first thing I noticed was that at any given moment, most of the troop was inert—not moving. Usually only one jay was ever shifting perches at any time. Intrigued, I made it a point to keep an eye on each and every bird. Noting their positions. Watching to see if ever more than one bird was changing position.

It wasn't hard to do. In a black and white landscape, the all-blue birds stood out.

As it happened, there were one or two times that, it seemed, somebody jumped the gun—i.e., that two birds shifted position simultaneously (or, in actuality, one bird made its move before another had settled). But as a general rule, only one bird was moving while the other members of the group were watching—and very commonly the bird that was moving was foraging. Staying low. Descending to the forest floor where acorns were the prize.

It was about this time that I realized that not only were the movements deliberate, so too was the positioning. Some birds were high. Some were low. One or more were ahead of the foraging bird. One or more behind. I don't recall seeing the foraging bird out ahead of the group, or behind.

The foraging bird was in essence within a protected perimeter that maintained its defensive configuration as the flock advanced— and perhaps an offensive configuration, too. Perhaps the advance birds were searching for food to exploit. The deployment was as well suited for search and find as search and survive.

And of course all of the birds, whether foraging or standing post, gained a measure of protection by having other members of the troop watching for danger.

I don't know how discipline was maintained. I don't know how birds knew their assignments or how duties changed or were shared (because it's plain that every member of the troop needed, at some point, to feed).

I, after all, was stuck in my tree. The birds had a whole woodland to exploit and moved on. My total observation time was less than two hours.

But the incredible thing, the marvelous thing, was had you asked me last November whether I was "familiar with Blue Jays," I would have said . . .

"Sure. Been watching them all my life." And I have! I've just never studied them. There's a difference.

A Bolt from the Blue

It was Thursday, three days after I saw the distant jay and formulated my questions, that I began to piece together the foraging pattern of the jays, and on Friday, from the same tree stand, I got to test it. To see if in fact the birds evidenced the same pattern and movements once again.

They did.

It was easy to keep the members of the troop in sight. The birds helped by staying, for the most part, in the open, on strategic perches. The stark woodlands helped too, serving as a cone cell–neutralizing backdrop so that the electric-blue birds stood out.

Even when the jays landed on the other side of some lattice-work of branches, their bright blue color simply cut through the clutter.

That's when it hit me like a bolt from the blue. That's when I realized why jays are blue—or at least one reason why jays may hold their flashy plumage year-round. It is precisely to stand out! It is, in the lingo of frontline troops and law enforcement personnel, for "situational awareness." It is so every bird can know where the other members of the troop are at any moment!

A single all-blue bird in a stark winter landscape is vulnerable. But with a disciplined troop of birds coordinating their movements, the eye-catching color confers a strategic advantage.

And supporting this supposition (because, untested, that is all it is) is the silence evidenced by my foraging jays. They could have communicated their whereabouts to each other vocally (as foraging chickadees, kinglets, and nuthatches do). But they did not. They foraged, for the most part, in silence. The only thing loud about them was their plumage.

The Winter Blues

It's amazing, really, the things you can learn about a bird you've "known" your whole life just by spending time watching. Sitting here now, writing this sentence, it makes me wonder how much I don't know about all those other birds I've "known" all my life.

And what might be learned. One thing I will say: If anyone in the Pentagon is trying to figure out how to move a squad through hostile territory and optimize its offensive and defensive capacity in a three-dimensional world, they might want to consider hiring a team of ornithologists to study Blue Jays—learn how the birds in blue do it.

And now that I have the situational awareness to understand that there are lots of things to discover about the birds I already know, I'm planning to spend more time bird watching and less time bird shopping.

Like this morning, when I went to check out the feeders here at the Cape May Bird Observatory. There were three Blue Jays perched in the bush between me and the feeders. One was looking left. One was looking right. One was watching straight ahead. The building was guarding their rear. The birds held this defensive position until I left.

Ha! Probably would have missed the significance of this back when I was just *familiar* with Blue Jays.

Fifteen Things to Do With Your Stockpile of Retired Binoculars

Experienced birders differ in many respects. Some well-traveled types keep extensive Life Lists; others have lists that are more modest. Some are plumage driven in their approach to bird identification; some are more attuned to bird shape. But there is one thing that experienced birders all seem to have in common. They own a stockpile of used binoculars accumulated as they climbed the glass ladder en route to the alpha binoculars they brandish today.

Whether those old bins are hidden in a trunk beneath a high-school letter sweater or hanging on a hook in the closet where holiday decorations are stored, chances are you have given thought to finding a use for them (if for no other reason than to keep you from seeing them and recalling how much money you've wasted as you climbed that glass ladder).

So here, for review and consideration, are fifteen things you can do with old binoculars.

1. Take them apart. Try putting them back together so you will finally appreciate why you have been repeatedly told "never, never, never try to repair misaligned binoculars yourself."

2. Put them in the trunk of your car in case you have a flat on a hill and need something to chuck behind the wheel, or, in the event that you slide off a snow-covered road en route to

a Christmas Bird Count (and you don't have a set of chains), you can drop them in the snow for traction.

3. Plop one (or two or three) in the water-holding tanks of your toilets to conserve water.

4. Use them as walking weights to increase upper arm and shoulder strength for the upcoming hawk-watching season or spring warbler migration.

5. Wait until the leaves are about to fall, then stuff them into the drainpipes of annoying neighbors who fire up their lawn mowers at seven on Sunday mornings (not that you are around to hear it).

6. Post a listing on eBay. Promote them as one-of-a-kind instruments given to you by aliens in the desert outside Roswell, New Mexico, that allow you to communicate with advanced civilizations in other worlds, see Elvis, eat all you want and still lose weight, and view tomorrow's stock closings. Keep posting until you get rid of the lot.

7. Put them on the roof of your car and drive away (because everyone does this at least once in their life, and wouldn't you rather do it with a glass you don't want?).

8. Put them under your pillow in hopes that the Optics Fairy will exchange them for one of the hot new roof prisms on the market. If not the Optics Fairy, then maybe the Tooth Fairy will be making the rounds and exchange them for a quarter. Either way, you're ahead.

9. Try to convince Tom Sawyer to take them in exchange for a turn at whitewashing Aunt Polly's fence. If this fails (and chances are it will—Tom's a pretty smart boy), track down the kid with the dead rat tied to a string (used, if memory serves, to swing at girls to keep them away). Trade the binoculars for the rat, then trade up for the paintbrush and a place at the fence (or keep the rat).

10. Hide them in obscure places all around your house so that you can practice finding them in preparation for the next

time you're off on a birding trip and can't find your alpha binoculars.

11. Bury them on beaches so that people wandering around with metal detectors will actually have something to find.

12. Mail them to someone you don't like (postage due) and put the name and address of someone you like even less in the return address slot.

13. When walking through enchanted forests, drop them to mark your path so you can find your way home. Save the bread crumbs for chumming pelagics.

14. And when you take that pelagic, and if you have binoculars left over, consider the merits of building a reef.

But the best thing to do with used binoculars in good working condition is, of course:

15. DONATE THEM TO SOME NEEDY YOUNG BIRDER. Not only will they serve as a means to propel a youngster into a world of wonder and discovery, they will serve as a seed from which will sprout a bottom drawer of binoculars that they in turn can pass on.

I Say What You Mean

Birding is a subculture (abbreviates to "cult"), and like most subcultures, we have our own standards, rituals, and, most importantly, language. While English is most certainly the foundation of birderspeak, it digresses widely from this commonly held ground. Beginning birders (and nonbirders) are often confused by the gulf that exists between what birders say and what birders mean.

Here, for quick reference, and with the aim of clarification, are rough translations of what birders really mean when they say what they say.

WHAT BIRDERS SAY: "Oh, I'm not a Lister."

WHAT THEY MEAN: "My Life List is fewer than 300 species."

SAY: "I keep track of the birds I see. But I wouldn't call myself a Lister."

MEAN: "My Life List is approaching 500."

SAY: "Sure, I'm a Lister."

MEAN: "My Life List just topped 600. Ha!"

SAY: "You bet I'm a Lister."

MEAN: "I'm closing in on 700 species."

SAY: "I got a quick look at the bird. Not enough to satisfy, but more than enough to clinch the identification."

MEAN: "I got a fleeting glimpse of a bird going dead away, into the sun. Somebody else said it was the Whiskered Tern."

SAY: "I'm pretty sure that's what it was, but I'm going to check some plates tonight to be certain."

MEAN: "Bird's good as got a checkmark in front of its name."

SAY: "The bird was here a minute ago. You just missed it."

MEAN: "Bird hasn't been seen for at least twenty minutes."

SAY: "You should have been here five minutes ago."

MEAN: "The bird hasn't been seen for over an hour."

SAY: "Close? The bird was right here."

MEAN: "Two football fields away."

SAY: "Oh, really close."

MEAN: "The length of an average jet runway."

SAY: "Reasonably close."

MEAN: "Bird was a speck."

SAY: "Pretty far."

MEAN: "It was an upended soda bottle half buried in the mud, two time zones away, made animate by heat waves."

SAY: "Pretty far, and the angle was poor."

MEAN: "It was a soda bottle lying on its side and half buried in the mud."

SAY: "I've had these binoculars for thirty years and wouldn't part with them for anything."

MEAN: "I'm considering an upgrade and trying to justify the expense."

SAY (after looking through somebody else's new HD spotting scope): "Well, I don't see a lot of difference between yours and mine—not for the money."

MEAN: "Damn! Why didn't I just bite the bullet and spend the extra money for the superior glass?"

SAY: "Well, what I enjoy most about birding is seeing the birds of my region."

MEAN: "I have alimony payments, a new spouse, new mortgage, two cars bought on time, and there're rumors of cutbacks in my department. Trip to Ecuador? Not until the kids are out of college."

SAY: "For my money, there's never been a better field guide than the good old . . ."

MEAN: "I'm more or less locked into the wing bars and tail spots age of birding."

SAY: "I know Sharp-shinned Hawks, and if that was a Sharp-shinned Hawk, then it has to be the biggest, rangiest, most . . ."

MEAN: "OK, you win. It was Sharp-shinned."

SAY: "My ears have gotten so bad these last few years that I have to relearn warbler songs every spring."

MIGHT HAVE ADDED (BUT DIDN'T): "But then, I always did."

SAY: "No problem. The bird is everywhere."

MEAN: "I've seen one or two around there in my life."

SAY: "Oh, the bird's there all right. But it takes time to dig one out."

MEAN: "You haven't got a prayer of seeing Black Rail."

The Star

In front of the auditorium the bird club's officers were up to their rolled-up shirtsleeves in matters of state—trying to decide whether to serve cardboard beef or rubber chicken at the annual banquet. Next item on the agenda: fingering a new editor for the club's newsletter.

In the aisles, the club's rank and file were gathered into conversational clots. Some were discussing their most recent package tours to Alaska (and Belize . . . and India . . .). Some were arguing the merits of fixed-power magnification on a spotting scope versus zoom. Others were taking advantage of another member's absence to raise questions (and eyebrows) about the "reported" Clark's Grebe.

Standing away from the crowd, leaning against the wall, was a man with eyes that were alert and a face that was shy. He was more younger than older. He was neatly dressed in the way that people who don't have lots of money take pains to be presentable. He was responding to a question from a woman who wanted to know how to tell Purple Finch from House Finch . . . who was replaced by a gentleman who wondered whether it was possible for him to have seen an Anhinga . . . who was interrupted by a man who wanted directions to the reported Little Gull . . .

He was the club's field-trip chairman. He was the quiet star. He'd never been to Alaska (or Belize or India). And although his Life

List was not very impressive, not even close to the numbers espoused by some of the well-traveled officers of the club, nobody had seen more birds in the county than he, and nobody knew more about their habits.

When they hold the roundup for the annual Christmas Bird Count and the compiler pleads "Saw-whet Owl?" all eyes turn reflexively to him.

When plans for a corporate park are proposed and nobody knows whether there are any threatened or endangered species on the site, the person the town's environmental committee calls is . . . well, you guess.

And during the meeting—after the beef-firsters have had their way, after a brand-new editor has bravely accepted her charge, after the club president has made the call for "recent bird sightings," and after everyone else has had their say—he is the one who offers a five-minute accounting of avian encounters that carries across a hushed room.

He is their champion. The best among them. His achievements become theirs.

After the meeting, after the members have all gone home, he walks to the woodlot behind the darkened meeting hall and starts to whistle. It's a quavering whistle, winsome too, and it doesn't take two minutes before a screech-owl responds. They continue their conversation, bird and birder, for several minutes, then go their separate ways. It was the only conversation he'd initiated all evening.

Why Good Gulls Go Bad

As Guardian of the Mystic Swizzle Stick of Alexander Wilson and Holder of the High Bum-Wipe for the Nth Order of the Avian Vent, I am often called upon to address weighty bird-related questions that confound science and vault the ages.

Most recently I was asked why "seagulls" steal french fries. First, you should understand (as I have pointedly pointed out before) that "seagull" is a misnomer. Insofar as most gulls are found within fifty miles of seacoasts (not well offshore, like shearwaters and other tubenoses), this bird group might more accurately be called "coast gulls" or "land gulls."

What's a tubenose? Sorry. One question per essay.

Anyway, seagulls steal french fries because you let them. You let them because you are an inept guardian of your food. By being inept, you invoke the kleptoparasitic instincts of gulls.

Klepto—to steal; *parasitism*—the act of dining at the expense of others. While this definition, strictly speaking, does not exclude visiting houseguests, in avian terms it is most commonly applied to describe the feeding behavior of birds that make a good living pilfering the food obtained by other birds and animals.

You've seen this behavior on the Nature Channel. Every other evening, it airs footage of Grizzly Bears scooping up salmon on rivers in Alaska. You will note that there are always lots of gulls standing around. The gulls don't catch the fish. The gulls are waiting

to cash in on what the bears leave behind (this is called scavenging), but if the gull rushes in and grabs morsels right out from under the bruin's nose—that's kleptoparasitism.

Salmon is to bear is to gull what french fry is to tourist is to gull.

You drop fries on the boardwalk, they belong to the gulls. You get careless with your fries, you lose them to a gull.

The problem of kleptoparasitism on the part of gulls is exacerbated by misdirected philanthropy. Some people (most commonly tourists) think it's fun or cute or beneficial to take philanthropy to an interspecific level and offer food to gulls—an act in league with teaching a homicidal maniac how to speed-load a revolver.

Well, maybe not quite *that* extreme.

Actually, the most aggressive display of kleptoparasitism I've ever experienced happened not on the boardwalk of Cape May but in a parking lot in Denali National Park, in Alaska. The marauding larids were Mew Gulls, not Laughing Gulls, and how I came to lose my ham sandwich was the result of my raising binoculars to my eyes with my right hand while attempting to steady the glass with a sandwich-clutching left.

What this means is that I raised a sandwich in the air—just like all the stupid tourists who have honed the pilfering skills of gulls by offering them food in just this manner.

The gull did its part and scored. Ripped the entire sandwich right out of my unsuspecting hand. Some misguided tourist gets credited with an assist.

So don't help a good gull go bad. Restrain your philanthropic instincts. Remember, it starts with french fries, escalates to ham sandwiches, and then . . .

Well, I don't want to start a panic, but I'm pretty sure I saw a couple of Laughing Gulls casing two of Cape May's finer restaurants the other day.

Announcing CertiBird

Attention all birders with ten fingers and toes, or anyone who has ever had or knows someone who has ever had their bird sighting(s) rejected by a state bird records committee:

Announcing CertiBird.

Yes, CertiBird. A verification service that guarantees your sighting will receive the most considered regard and highest level of affirmation your credit limit can accommodate.

It doesn't matter if your ability to conjure rare and unusual birds borders on the hallucinogenic or that your name attached to a rare-bird sighting report makes state records committee members curl into the fetal position, put their thumbs in their mouths, and rock back and forth.

With CertiBird, no matter how outrageous your observation, no matter how little documentation you can produce, your sighting cannot be rejected. Once payment clears, certification is guaranteed.

You can choose from one of several levels of affirmation—Kingdom, Phylum, Class, Order, Family, Genus, Species, even Subspecies. In most cases, your CertiBird Certification will cost you less than the costs that would be incurred by traveling to a species' established range and actually seeing the bird.

Your CertiBird Sighting Certificate, suitable for framing, guaranteed not to lose credence, authenticated by the CertiBird governing board and signed by Alexander Wilson, John James Audubon, and P. T. Barnum, will be sent to you and a copy forwarded to your state records committee. In addition, your record will be automatically and permanently deposited in the CertiBird databank, where it will accrue additional credence at a fixed rate of three percent per year.

Compare this to other plans that merely accept or reject your sighting or, in most cases, ask for "additional supporting evidence," leaving your sighting in limbo for months, sometimes years.

And with the CertiBird Straw Into Gold Plan all your sightings are preapproved. Any new species you discover will be named after you, your loved ones, or anyone you choose to honor, then added to the CertiBird Checklist of Intragalactic Species—the most comprehensive list of bird species in the universe.

Many people believe that finding a bird species new to science is a remote possibility. The truth is that many CertiBird subscribers routinely encounter species whose descriptions do not resemble any known species. Some plan subscribers find two or three new species a month.

Having the CertiBird Plan means you will never go birding dreading the possibility of finding something rare or unusual. And you will never suffer the embarrassment of having your sighting reviewed and rejected by committee members whose skills and knowledge of a region's avifauna differ from yours only by degrees.

Perhaps you think that your records committee will change its mind and accept your reported Labrador Duck. Maybe you're hoping that your good nature, enthusiasm, and additional information (to support the 400 pages of documentation and 120 color photos of lily pads marked with grease pencil arrows noting where the duck went down) will make a difference. The fact is, records committees rarely reverse their decisions, and then only when supporting evidence warrants such a reversal.

CertiBird guarantees that your sighting will be accepted. And if you choose to contest the decision of your records committee, the CertiBird Legal Team of Diogenes, Hammurabi, Solomon, Sir

Galahad, George Washington, Daniel Webster, the Oracle of Delphi, and Pinocchio are there to help.

There will never be a better time to enjoy the security of the CertiBird Plan. Rates will never be lower, and your subscription can never be canceled (so long as you keep making payments).

Every birder deserves the confidence of the CertiBird promise. And doesn't "every birder" include you?

Remember: When records committees ask "why us?" Certibird says "why not?"

Operators are standing by. Call now: 1-800-IT'S-MUTE.

Fuddy-Duddy Declaration

To Whom It May Concern (likely, nobody):

At 1:47 P.M., January 30, 2010, I officially declared myself a Fuddy-Duddy. I have seen the future of birding. I'm getting off at the next stop.

What inspired this epiphany? I recently found myself on a panel of "experts" for a bird-photo quiz. Three genuine experts, plus me, and a series of supernatural in-your-face images of birds projected on a screen. Our job: Name them.

It was fascinating. It was illuminating. It showed me that birding has reached a fork in the road. The future of bird identification is going one way. I'm going mine. But let's look back before jumping ahead.

About a hundred-plus years ago, birds were identified in the hand. The tool of interaction was the shotgun. You dusted the bird with shot. Picked it up. Noted distinguishing characteristics (length of the tarsus, color of the inside of the mouth) and identified the bird.

After the fact, of course. It wasn't real-time observation. It wasn't even field identification. It was just bird identification.

Then we got a new tool. Binoculars. They did what the shotgun did (provide supernatural intimacy with birds) but it happened in real time, with real live birds, and it spawned the age of "field identification."

For over a hundred years, we've been getting better and better at identifying birds in the field based on real-time study using binoculars.

Your field guides are based on this principle, as are your acquired skills.

Then I sat on the panel. Was shown a frozen, supersharp image of an adult accipiter in flight. Sharp-shinned or Cooper's? Said, "Cooper's." Moderator said, "Sharp-shinned."

I asked, "Why?" Answer: "Dark shadow on the face curving under the eye."

I said, "Huh?" Seen about two million accipiters—ninety-nine point ninety-nine percent in flight. Never noticed a five o'clock shadow on an adult Sharp-shinned.

Why so undiscerning? Because to apprehend this subtle feature, you have to freeze the bird in an image.

In essence, you have to "collect it." No more real time. Shoot first. Identify later.

Read on.

Five years ago, I was coleading a trip to the Antarctic. We had a swarm of prions in front of us. Question was, which species? Because of distance, animation, the limits of optics, and human perception, we couldn't apprehend enough of the face pattern to be certain.

A photo was taken. Enlarged. Result? Perfectly identifiable Antarctic Prions. The camera captured what the eye could not.

So my problem is that digital photography has given bird identification a new edge? No. I can live with that (even though we're back to collecting). My problem is where I see this new technology inexorably heading.

Consider: There are imaging systems that can read a bar code on the windshield of a car as it speeds past a tollbooth. There are programs that allow passenger screeners at airports to identify individuals by scanning their eyes.

How long do you think it's going to be before there will be bird identification software, satellite-linked to your techno-binos, that will capture the image of the bird you see? Download the image to your computer at home. Identify it. File it in your Day List, Year List, County List, Life List . . . all for later review.

And one hundred years of compounded and refined field identification skills become about as useful as knowing how to calculate a square root on a slide rule or change a typewriter ribbon.

So I have, accordingly and cheerfully, proclaimed myself a Fuddy-Duddy. A stone-knives-and-binocular-toting birder who is content to identify—and misidentify—real birds in real time in the field. Identifying frozen images of birds after the fact just doesn't interest me.

Funny thing about that accipiter (whatever it was). If it had been a real bird instead of a frozen image, it would have been moving, flying—perfectly identifiable if studied by any competent hawk watcher.

For now, anyway.

Chipper

It comes as something of a surprise to me that I have not written this essay before. Or maybe I did—many years ago when the memory was still fresh and raw—but misplaced it.

Events that change lives deserve more and better. That one week in the eighth summer of my life taught me more about loss than any other that followed.

At least any so far.

And I also learned a great deal about human unmindfulness and how well-intentioned actions have unforeseen consequences. Only later did I come to learn that the little personal tragedy that occurred at 57 Washington Avenue was being played out all over the world.

Others have spoken about this, the big picture, including and most notably Rachel Carson. I'm here to tell you about Chipper, a robin nestling that was, according to the testimony given to my mother, "abandoned" into my care.

It's a drama that plays itself out all across suburbia every summer. A child (nestlings themselves) finds baby bird on the lawn. Captures it. Cradles it. And a bond is struck.

Two minutes later some parent is confronted by a fistful of feathers and a look so imploring it neutralizes logic and reason.

"Put the bird back where they found it," I've counseled a thousand parents in my time. From late May into August, phone lines at nature centers across the planet are jammed with BBCs (baby bird calls). Some parents are seeking backup for their well-considered inclinations—an authoritative voice to support their determination to let nature take its natural course. But some parents have already been turned, won over by the beady-eyed wonder wrapped in their child's hands, and their child's imploring look.

My mom fell into the latter category. And Chipper found his way into a large box filled with newspaper. For the next week, the center of my universe was framed by the open bill of a demanding young robin.

Not being an adult robin, I didn't know a great deal about the care and feeding of young robins. I knew they ate worms. I thought they needed water—which is not the case and explains why the newspaper on the bottom of the box had to be changed so often. Real American Robin parents don't have eyedroppers at their disposal, and young robins draw all the moisture they require from their food.

But worms were something I was pretty good at finding. Fishing was an almost-every-day activity, and worms were as essential to fishing (and childhood) as a bobber, split shot, and hook. The problem and challenge was that Chipper arrived in a summer marked by severe drought. Worms were deep. Even professional robins were challenged to find them.

I must have turned over a tetherball court's worth of planet in my search of worms that were coiled in balls deep in the soil. My hands were blistered, and there was no one to tell me that when it comes to feeding nestlings, canned dog food works just fine.

It didn't matter. Not the blisters, not the hours away from TV, not the sandlot baseball games I was missing. The fruits of my labor blossomed every time I held a tidbit of worm in front of Chipper's mouth and watched it go all bright yellow and needy. Chipper was the very first creature that needed me, depended upon me. I took my obligation seriously.

Of course I also took a few liberties with my new charge. "Taught" the bird to perch on my fingers. Further "trained" him to ride on my shoulder. Why he was out of the nest in the first place?

I don't know. Maybe there had been a storm. Maybe there were too many nestlings and Chipper got too adventurous and was crowded out. Likewise, and unlike most nestlings that are "abandoned," no adult birds seemed to be attending to Chipper before I carried him home.

I actually did wait. Waited and hoped. He really *did* seem to be orphaned, which was almost certainly not the case.

I did realize, of course, and right from the get-go, that it was Chipper's destiny to fly. Worms and feces I could handle. Teaching a young robin to fly was a problem that troubled me greatly.

I secretly hoped that some adult robin would swoop in and offer Chipper some rudimentary flight training. I was at a point where I was leaving Chipper outside, on a specially made perch, so he could socialize with other robins—his peers. Learn the language. Get some pointers on takeoffs and landings.

It broke my heart to think that someday Chipper would fly away. Migrate south. But, well, it's something that all parents must deal with, isn't it? That endgame for all our nurturing care is independence. Hard truth.

As it turned out, I didn't have to worry about flight lessons or even bon voyage. Life had a harder lesson in store.

I don't know why I left the worms out on the front porch that night beneath a shallow layer of dirt. I didn't give them a first (or a second) thought when the DDT-fogging truck rolled past our house, filling the neighborhood with its killing white cloud.

The stuff was harmless, we were told. Killed mosquitos, a good thing.

The next morning, as soon as Chipper's bill blossomed, I satisfied him with as many worms as he wanted—which proved to be more than enough. A lethal dose. It took several hours for Chipper to convulse his way to peace. I'm only glad I didn't know then how nerve gas kills.

I only know that I was smart enough to put two and two together. Link the spray to the death. And while the causal link was me, the problem was the spray.

The adults were wrong—this stuff killed birds, not just bugs!

So I wrote to the mayor of Hanover Township and told him. Told him about Chipper. Told him to stop the spraying.

Guess what? They didn't pay any attention and they didn't stop. What the heck did I know? I was just a kid.

But I was in good company. Two years later, a heroic biologist put her career on the line and told the world, in a book called *Silent Spring*, that DDT was poisoning the planet. Initially, they ignored her too.

And while Chipper's story probably didn't make the senatorial hearings about the danger of DDT, there were plenty of other Chippers—plenty of evidence that our species had introduced a substance that was detrimental to the environment and . . .

In the name of sanity, we had to stop using it. Ultimately, in North America, we did. It has taken more than thirty years for the imbalance to the bird world caused by DDT to be righted. Kind of makes you wonder what's going on out there right now, doesn't it?

Chipper? He ended up in a shoebox, buried in the backyard, the fate of many young birds taken in by kindness. Human kindness is many things, but it is an imperfect substitute for nature.

My take-home? My lesson? That nothing done to, even for, the natural world is done with impunity. We engage it with risk but no promises. Only lessons and memories.

Whence Came Caps

Note to readers: This essay/conversation is not verbatim, but it does capture the essence and elements of my discussion with my friend Paul Baichich.

I received a call the other day from friend and ornithohistorian Paul Baichich. Paul is to bird study what Talmud scholars are to Torah interpretation, and at the heart of almost every call is a question. This call was no exception.

"When did baseball caps become the rage among bird watchers?" asked the Sage of Aves.

Not being one of the beautiful birders, I've never paid a great deal of attention to birding fashion trends. Besides, I hate wearing hats.

"Sorry, Paul. Couldn't say. I'm a visor man myself."

"Well, isn't a visor just a baseball cap with the top cut off?"

"It is not," I bristled. "It's a utilitarian masterpiece. An honored badge of the Order of Hawk Watchers. Why are you so curious about baseball caps?"

"Well, Wayne Peterson and I were touring with a bunch of European birders a few years ago, and it occurred to us that we were the only ones wearing baseball caps."

If this observation doesn't sound like fertile ground for a question, or an essay, then you don't know Paul. Or me. Frankly, the two of us just go gaga over this kind of avocational trivia.

"You don't suppose it has something to do with baseball being an American institution?" I invited.

"Of course it has something to do with baseball and *E Pluribus Unum!*" Paul chastised. "But baseball caps have been around since Garfield."

"The president or the cat?"

"President," he said. "Modern birding goes back almost as far. To the late 1880s. I defy you to find a photo of Chester Reed or Frank M. Chapman wearing a New York Yankees cap."

I had to admit it would be challenging. Particularly since Chester Reed was almost certainly a Red Sox fan. But Paul's question was nevertheless apt. For most of birding's first century, birders went bareheaded or wore fedoras. Big wide-brimmed affairs that fell prey to every modest gust of wind. What winds of change could have prompted such a dramatic fashion shift? And when?

"How about the 1940s?" I suggested. "About the time birders started heading up to the North Lookout at Hawk Mountain? I'll bet there's a pyramid of wind-tossed fedoras lying below. A tight-fitting baseball cap would save many a trip down to the River of Rocks."

"Thought of that," said Paul. "I looked in *Hawks Aloft* and *Birds Over America*. Fedoras were still the rage—except, of course, for Rosalie Edge's upscale headpiece."

"Something with a plume?" I suggested.

"I think Rosalie Edge was too smart to troll for goshawks," Paul assured.

"Right," I agreed. "So what's your guess about baseball caps and birding?"

"I seem to recall the trend started in the seventies or perhaps early eighties," said Paul. "There's a photo of Roger Peterson wearing what looks like a baseball cap on the 1983 California Big Day."

I glanced over at my office wall. At the photo of Roger sent to me in 1985. I tried to picture that celebrated mane of white hair sticking out, Ronald McDonald fashion, from under the brim of a baseball cap.

"Are you sure?"

"But in Arthur Allen's *Stalking Birds with Color Camera* there's a photo of the 1950 National Audubon Society convention field trip to Long Island showing four baseball-type caps in the crowd—none with logos."

Four. A similar photo taken today would show four hundred. Heck, there isn't a whistle-stop bird club, crossroads bird festival, or hole-in-the-wall nature center that doesn't have an emblematic baseball cap for sale. They are badges of our prized affiliations, tokens of our travels. You go to the Rio Grande Valley Birding Festival or Cape May and you want to flaunt your affiliations to all your birding friends . . .

"Attu?!" I suggested.

"Gesundheit," Paul replied.

"No. Attu. Island. Coast Guard LORAN station. Baseball caps are standard garb among Coasties. I'll bet every birder that went to Attu purchased an emblematic cap from LORAN Station Attu. And didn't birders start going to Attu in the late seventies or eighties? That fits your timetable."

"You're suggesting that the United States Coast Guard is responsible for birders wearing baseball caps?"

"It's just a theory. And it also explains why European birders don't wear baseball caps."

"Why's that?"

"Old World birders aren't interested in going to Attu to see Old World birds. And it's a long way to go just to buy a hat. Now I've got a question for you. What about birder vests? When did jackets without sleeves become birding apparel, and just how many pockets does a person need to carry a field guide and a notepad?"

We never did come up with a really good answer about baseball caps and birders. We're still working on the birding vest question, too.